W9-BNO-419

Hartland Public Library
P.O. Box 137
Hartland, VT 05048-0137

Richard Hamblyn

FARRAR, STRAUS AND GIROUX
NEW YORK

The Invention of Clouds

The Invention of Clouds

How an Amateur Meteorologist Forged the Language of the Skies

Farrar, Straus and Giroux
19 Union Square West, New York 10003

Printed in the United States of America
First edition, 2001

Library of Congress Cataloging-in-Publication Data

Hamblyn, Richard, 1965–
The invention of clouds : how an amateur meteorologist forged the language of the skies / Richard Hamblyn.— 1st ed.
p. cm.
Includes index.
ISBN 0-374-17715-5 (hardcover : alk. paper)
1. Clouds. 2. Meteorology—England—History—19th century. 3. Howard, Luke, 1772–1864. 4. Meteorologists—England—Biography. I. Title.

QC921 .H35 2001
551.57'6—dc21

00-068189

Designed by Lynn Buckley

Grateful acknowledgment is made for permission to reproduce the following:
Fig. 1: Cirrus, in the engraved version of Luke Howard's original drawing, from Luke Howard, *On the Modifications of Clouds*, 1804, by permission of the British Library, BL 1393.k.16.(1). Fig. 2: The street cloud seen by Hayman Rooke from the window of his study, from Hayman Rooke, *A Continuation of the Annual Meteorological Register, kept at Mansfield Woodhouse*, 1802, by permission of the British Library, BL 08756.b.49.(2.). Figs. 3 and 4: The Plough Court laboratory in a nineteenth-century photograph, c. 1860, and No. 2 Plough Court, Lombard Street, c. 1840, reproduced with kind permission of Glaxo Wellcome plc, Greenford, U.K. Fig. 6: "A Balloon Prospect from Above the Clouds," from Thomas Baldwin, *Airopaidia*, 1786, by permission of the British Library, BL 1137.c.17. Figs. 11, 12, 13, 14, 15, 16, and 17: Cirrus, cumulus, and stratus; Cirro-cumulus, cirro-stratus, and cumulo-stratus; Nimbus, from Luke Howard, *On the Modifications of Clouds*, 1804, by permission of the British Library, BL 1393.k.16.(1.). Fig. 20: Watercolor of clouds by Luke Howard, c. 1807, by permission of the Science and Society Picture Library, U.K. Fig. 21: Portrait of Luke Howard by John Opie, c. 1807, by permission of the Royal Meteorological Society, U.K. Fig. 22: The original Beaufort scale of 1806, Crown Copyright. Fig. 27: Photograph of cumulus cloud by Ralph Abercromby, from *The International Cloud Atlas*, 1896, by permission of the British Library, BL 8753.dd.17.

For Jo Lynch

Contents

List of Illustrations

The Invention of Clouds

Prologue

The Useless Pursuit of Shadows

"Then what do you love, you extraordinary stranger?"
"I love clouds . . . drifting clouds . . . there . . . over there . . . marvelous clouds."

Charles Baudelaire, 1862[1]

At six o'clock one evening in December 1802, in a dank and cavernous laboratory in London, an unknown young amateur meteorologist untied a bundle of handwritten pages, carefully balanced a roll of watercolor drawings beside his chair, and prepared himself to speak on a subject curiously at odds with his subterranean surroundings. It was a cold evening, colder still in the basement of the old building in Plough Court, and as the young man rose to address his audience, answering the supportive smiles of one or two of his friends, his slight shiver might have been due to the cold as much as to anticipation.

He was dressed simply, in an unadorned black coat and a high white collar—the young urban Dissenter's badge of plainness—and his reticent demeanor spoke of a natural modesty as well as trepidation. He could never, of course, have imagined that the evening was to make him famous, and as he cleared his throat and stared at the title of his lecture, "On the Modifications of Clouds," there was nothing in the air to suggest that his life was about to change.

The usual discomforts of public speaking would have been worse for a Quaker, and worse still for one as self-doubting and preoccupied as the 30-year-old chemist Luke Howard. Howard knew that his talents were not of the incendiary kind. They were not those of the flamboyant young Cornishman Humphry Davy, for example, whom he had recently met and whose rising fame as a scientific speaker told loudly of the worldly rewards of masculine looks and self-assurance. But Howard, whose mild hazel eyes peered out from his slender and serious face, at least felt himself to be largely among friends. Perhaps it would strike him later how unlikely his situation looked as a candidate for legend: there he was, an unknown speaker in an inauspicious room, the very subject of whose talk was new and untested. The subject was so new, indeed, that it had no defining term. Depending on how his ideas were received, the study of clouds might be hailed by his audience as a new and necessary branch of natural philosophy. Or if things went wrong that evening, as he suspected they might, the enterprise itself might be dismissed in its entirety as a useless pursuit of shadows.

Most pioneers are at the mercy of doubt at the beginning, whether of their worth, of their theories, or of the whole enigmatic field in which they labor. Luke Howard was no exception. His hesitations, however, were beginning to attract the notice of the room. He registered an expectant silence among his audience, and someone from the blank of faces nodded at him to start. Some of the older audience members and their guests, after all, had to be over at Somerset House by eight o'clock that same evening, for the start of the more prestigious meeting at the rooms of the Royal Society (and of course for the excellent three-course dinner that would be served to them afterward in the dining room). They would have been in no mood to find themselves delayed by the hesitations of an unknown amateur cloud watcher.

But little did they know how they would continue to speak of the evening before them for years to come, or how that coming hour would live so long and so vividly in their memories. For they had been there when Luke Howard spoke; they had been there at the unfolding of an epoch.

As unaware as the audience of what fortune held for his future, Luke Howard took a deep and steadying breath and, like a listener at a spoken recital, heard from afar his own quiet voice recounting the opening words of his address:

> My talk this evening concerns itself with what may strike some as an uncharacteristically impractical subject: it is concerned with the mod-

ifications of clouds. Since the increased attention which has been given to meteorology, the studies of the various appearances of water suspended in the atmosphere is become an interesting and even necessary branch of that pursuit. If Clouds were the mere result of the condensation of vapour in the masses of atmosphere which they occupy, if their variations were produced by the movements of the atmosphere alone, then indeed might the study of them be deemed a useless pursuit of shadows, an attempt to describe forms which, being the sport of winds, must be ever varying, and therefore not to be defined. But the case is not so with clouds . . . [2]

And as the hour wore on to the sound of Howard's voice, a singular journey began, a journey that would lift an unknown speaker from a chemical factory located in a courtyard off Lombard Street, EC1, up into the realms of scientific and romantic celebrity. It is an hour to be remembered by historians and daydreamers alike, for by the end of his lecture Luke Howard, by giving language to nature's most ineffable and prodigal forms, had squared an ancient and anxiogenic circle.

For by the end of his lecture Luke Howard had named the clouds.

Chapter One

The Theater of Science

Science, illuminating ray!
Fair mental beam, extend thy sway,
And shine from pole to pole!
From thy accumulated store,
O'er every mind thy riches pour,
Excite from low desires to soar,
And dignify the soul.

Sarah Hoare, 1831[1]

It might seem difficult to imagine now, in this era of cool detachment, but in the opening years of the nineteenth century people cheered loudly at lectures. While filing through the doors into a lamplit hall, upon the arrival of the speaker and his mercurial props, or at signal moments of disclosure and display, audiences found opportunities to make themselves heard. It mattered little whether the speaker was a mechanic, a meteor zealot, or simply an amateur showman on a mission to explain. Anyone with confidence and good vocal projection could arrange to appear at one of the endless assemblies of paying spectators that were springing up fast throughout the expanding cities of Europe and North America.

The full range of the philosophical shows and diversions available to audiences at the turn of the nineteenth century was various and impressive, particularly in the towns and cities of Britain, and especially in London, where there was nothing isolated or unusual about a lecture such as Howard's on the clouds. As evening fell, the crowds assembled and the revelations began to unfold. And what a cast of revelations they were: every animal, vegetable, and mineral known to man, samples of all four elements, and challenges to all six senses, not to mention machines, inventions, and novelties of every kind, were regularly paraded before the eyes of an astonished and insatiable public. There were demonstrations of fireworks, hydraulics, magnetics, and mathematics. There were machines to show the revolutions of the planets, the eruptions of volcanoes, or the hidden operations of the human heart. There was even a machine—dubbed the Eureka by its maker—for the production of Latin hexameters.[2] Even a dead language could be brought to life by the magnificent actions of a machine.

By the end of the eighteenth century the grip of rational entertainment had firmly secured itself on the public mind, and had done so because it served the equal, if novel, demands of pleasure, instruction, and imagination. Science had been on the rise for a century or more, and had now ascended to its height, where it drifted through the cultural atmosphere of the age. London, already blessed with the finest scientific and medical instrument makers in the world, was now the center not only for empirical measurement but also for conjectural pleasure.

Such pleasure was relentlessly pursued. According to an article in the *Observer* on July 27, 1806, for example, the exceptional thunderstorm that had occurred the Thursday before "afforded ample opportunity at the Theatre of Science, 97 Pall Mall, to Mr Hardie's talents in defence of his new Theory of Lightning." It certainly did. The evening's entertainment had begun with what were by then familiar galvanic experiments ("among them the generation of various solid bodies from a mixture of different transparent gases") but had gone on to culminate in a spectacular display of "meteors, aurora borealis, real lightning and other phenomena," all demonstrated as alleged supports for Mr. Hardie's curious theory—stubbornly maintained against the growing evidence to the contrary—that meteorological activity had nothing to do with electrical force.

More noteworthy than the hypothesis itself, perhaps, was the fact that well over a hundred people had paid to hear it; they were crammed into every inch of available space, with latecomers standing at the back. This was profitable entertainment at its best, delivering what the metropolitan audiences of late Georgian England wanted most of all to see and to hear: the revelations of a profligate nature.

Yet Hardie's "Theatre of Science" was only one of dozens of such popular and paying concerns. West End theaters like the Haymarket, the Lyceum, and the Duke of York's, as well as coffeehouses, taverns, and riverside pleasure gardens, were toured continually by philosophical showmen with their arrays of scientific and pseudoscientific displays. The efforts of itinerant lectures such as James Ferguson of Banffshire (1710–1776) or the great Adam Walker of Westmoreland (1731–1821) served to define the mainstream of public scientific understanding: uncritical, nonspecialist, and wide-ranging in its enthusiasm for the spiraling diversity of knowledge. Long queues formed outside Walker's astronomy lectures at the Haymarket Theatre, where he showcased his illuminated twenty-foot model of the giddily revolving planets. His lectures, every bit as vivid as his props, were enormously successful, and he was soon able to buy himself a house in Hanover Square from the proceeds. Walker was foremost in the train of self-made scientists who earned their living by subordinating new findings in chemistry, physics, and astronomy to the glorious reign of Spectacle, ushering onto the stage in rapid succession their hydraulic and hydrostatic machines, their Copernican models of the revolving solar system, their automatic chess players and other mechanical marvels, and their baroque optical chimeras, such as the cloud of eerie smoke that slowly cleared to reveal the ghastly guillotined head of Antoine Lavoisier, the celebrated but doomed eighteenth-century chemist and tax collector.

Lavoisier had been executed in 1794 by a Revolutionary Tribunal that was alleged to have declared, through the summing-up of the judge at his trial, that "the Republic has no need of scientists." Although it has taken the French two centuries to come to terms with this act of uncompromising barbarism ("It took them only an instant to cut off that head, but France may not produce another like it in a century," as Joseph Louis Lagrange was to comment in tears), in England the episode was quickly recruited as a cautionary tale to be told against the excesses of French revolutionary fervor.[3] The gorily modeled head of the decapitated chemist, part of the "Phantasmagoria" show held at the Lyceum Theatre during the summer months of 1802, was flourished both as a piece of entertainment and as a tribute to the freedoms of British research. Luke Howard and his circle of philosophical friends, drawn from the young men and women of Dissenting London, were not alone in revering Lavoisier as a tragic intellectual hero, cut down in the prime of a brilliant career.

This theater of knowledge was an important part of the climate of the Enlightenment era, an age often characterized as the Age of Reason, and it furnished the background to Luke Howard's lecture. The leading players, among whose number he was soon to take his place, were always the scientific performers, and, as in any other branch of dramatic performance, the reputations of the major stars commanded large audiences and commensurately generous rewards. The greatest of these performers, celebrated by lecturer and pretender alike, was the young Cornishman Humphry Davy (1778–1829), who became a wealthy London celebrity during the opening years of the nineteenth century. He was renowned for his extravagant and explosive demonstrations, for his speaking energy, and for the mesmerizing eye contact with which he held his audiences spellbound during the entire chemical proceedings on stage. Davy was the dark-haired, romantic son of a Penzance carpenter, and his looks and language were those of a poet, albeit a poet of enormous worldly ambition. Samuel Taylor Coleridge and Robert Southey maintained that Davy would have been a great poet had he not become a great chemist, while admiring women were heard to whisper that his eyes were made for something besides poring over crucibles.[4]

Here, then, was a man of great charisma, of star quality, as we might say today, although had his own scientific research not been serious enough to overtake his growing reputation as a speaker, he would now be one of the many mostly forgotten performers of the learned London stage. But at the outset of his London career (which was to culminate in the presidency of the Royal Society), his fame was as a showman rather than an innovator. His work at the Pneumatic Institute in Bristol during the 1790s had already won him a reputation as the world's most incandescent public speaker; so it was inevitable that his career would take him to London, and when he gave his inaugural public lecture at the newly opened Royal Institution of Great Britain on January 21, 1802, porters were needed to keep the impatient crowds in check. Albermarle Street, it was reported in the press, was blocked with carriages for an hour. The new lecture hall itself, with its gallery, pit, and slanting stage, was designed to give the elegant building as theatrical a feel as possible. Separate entrances had been designed to prevent the social classes from mingling, for as in other recent attractions, such as balloon and parachute ascents, where "Sweeps, Gemmen, and Ladies all scamper'd together," the Royal Institution was a popular (and populist) attraction.[5] From there, Humphry Davy, black-eyed, magnificent, and unstoppable, rose to become the presiding spirit of public science in Britain.

The handbills for his lectures promised electrical demonstrations, galvanic experiments, and semicontrolled explosions of gases, all of which attracted large, excited, and extremely vocal crowds who willingly paid their money to (in the language of the fairground) enter and be amazed. And it was with genuine amazement that Coleridge, after attending one of Davy's demonstrations, recorded how a sample of ether "burns bright indeed in the atmosphere, but o! how brightly whitely vividly beautiful in Oxygen gas," while the lecturer himself was just as dazzling: "every subject in Davy's mind has the principle of vitality. Living thoughts spring up like turf beneath his feet."[6]

Soon entire audiences were to find themselves as captivated as Coleridge. According to one of Davy's earliest biographers, "the sensation created by his first course of Lectures at the Institution, and the enthusiastic admiration which they obtained, is at this period scarcely to be imagined. Men of the first rank and talent—the literary and the scientific, the practical and the theoretical, bluestockings, and women of fashion, the old and the young, all crowded—eagerly crowded the lecture room. His youth, his simplicity, his natural eloquence, his chemical knowledge, his happy illustrations and well-conducted experiments, excited universal attention and unbounded applause. Compliments, invitations, and presents, were showered upon him in abundance from all quarters; his society was courted by all, and all appeared proud of his acquaintance."[7] Humphry Davy was the man of the moment, the Horatio Nelson of dry land, and Luke Howard was in the audience for a number of his demonstrations. Like others, he was moved to wonder at the unbounded energy of the man. It was as if the spirit of scientific inquiry itself had found expression in the person of the genius from Penzance.

Davy remained the most celebrated scientist in Britain for another two decades, until his death in Geneva at the age of 50 deprived the world of his talents. His end, much mourned among the learned circles of Europe, was almost certainly hastened by carbon monoxide poisoning, the result of a lifetime of hazardous experiments devised to determine the properties of gases. He would breathe them in until, as often as not, unconsciousness intervened. Waking, he would find himself slumped at his worktable, with burning lungs and an aching head. Humphry Davy's life of scientific self-sacrifice was rounded by a sadly fitting death.

But why should the theater of natural knowledge have gained such a hold on the popular imagination at the turn of the nineteenth century? Why should its audiences have lined up for so long and clamored so loudly for more, as they did at the

Royal Institution? The answer lay both in the novelty of the subject and in the state of general science education at the time. The vast majority of the population, whether educated or not, had simply never witnessed such processes as these: magnesium, flaring intensely, burning with a kind of stellar light that few could have imagined existed on earth, or sodium, first isolated by Davy himself, fizzing profanely in a container of water with a diabolic mineral energy. Metals that burned upon contact with air, or drab-looking powders, harmless on their own, that when mingled in a jar combusted suddenly and violently to produce billows of noxious gas. Phosphorus, with its white flame and searing heat, "the devil's element" (and not just because it was the thirteenth to be isolated), was offered up not only as a spectacle but also, alarmingly, as a medical marvel for the treatment of tuberculosis and gout.[8]

The discovery and application of such substances served the growing needs of industry and technology, and their public display was increasingly becoming an integral part of the process. New kinds of natural and material knowledge were taking their place within the wider cultural definition of the age. The secrecy of the alchemists was giving way to the high-profile publicity of the chemists and the physicists—the natural philosophers, as they termed themselves—who promised to uncover the secret properties of the natural world. And they, unlike the earlier alchemists, were delivering on their promises with aplomb.

Their confidence was bolstered by newly emerging frameworks for scientific thought that emphasized nature's long-term capacity for slow and steady change.

Alchemists had sought the secrets of sudden transformations that lay at the heart of the material world, but research in the growth areas of scientific inquiry, such as geology, with its various associated branches of volcanology, seismology, stratigraphy, and mineralogy, was beginning to make it clear that the gradual, unseen processes that had shaped the world and its objects remained ongoing into the present. Distant catastrophe was not the only model of geophysical formation. The earth was still changing as it always had, in only just perceptible increments. The terrifying new idea arose that entire stretches of landscape were continuing to rise and fall under the unseen pressures of the earth, while water, the most powerful of the elements because the most patient, was continuing to shape and reshape itself across the yielding portions of the earth. All that stands now, here in the present, will at some unknown point in the future be violently borne away. The universe will never cease its dance of change and mutability, and the processes by which it moves and convulses, whether gradual or sudden in their overall impact, were now to be considered as the true subjects of natural scientific inquiry.

It is not hard to see how the idea of natural changeability provided much of the conceptual background for the development of scientific meteorology. Clouds and weather, perhaps more than any other world phenomena, show clearly that there is no moment in nature when nothing can be said to be happening. As clouds race toward their own release from form, they are replenished by the mutable processes that created them. They drift, not into continuity, but into other, temporary states of being, all of which eventually decompose, to melt into the surrounding air. They rise and fall like vaporous civilizations, and the challenge to early meteorology was to reveal their hidden dynamics to our sight.

Yet meteorology is not an exact science. It is, rather, a search for narrative order among events governed not by laws alone but by the shapeless caprices of the atmosphere. Weather writes, erases, and rewrites itself upon the sky with the endless fluidity of language; and it is with language that we have sought throughout history to apprehend it. Since the sky has always been more read than measured, it has always been the province of words. Nothing has changed since Samuel Johnson complained in the middle of the eighteenth century that "when two Englishmen meet, their first talk is of the weather; they are in haste to tell each other, what they must already know, that it is hot or cold, bright or cloudy, windy or calm."[9] The weather, as Johnson so pertinently suggested, generates language more efficiently than it generates knowledge, for while it is always available and always with us, it is equally always unclear. That is why we need to talk it through.[10] And in lectures such as those given by the unparalleled Davy on his platform in the hall on Albermarle Street, or by the young Luke Howard, facing his audience in the chemical basement off Lombard Street, that was exactly what was happening. The world was being talked, and being shaped by that talk, into a new kind of natural order.

The age was one of the great ages of talk, an age that raised the art of conversation to the status of a public act. Language coursed through the events that were hosted at the London lecture halls, and at the dozens of similar venues across the rest of the country, impressing them with a lasting cultural and scientific imprint. The river of words was in flood with the chemical poetry of minerals and machines: *galvanism, latent heat, elective affinities, the steady state*. New words and new ideas circulated rapidly like a spoken currency among new and ready audiences.

Images from scientific discourse began to permeate the wider language in an unprecedented way. Who could resist peppering their talk with mesmerism, magnetism, lodestones, and longitude? Who, like the Bluestocking Elizabeth Montagu, could resist writing to a friend in 1761 to ask that she "make the same distinction between my heart, & those that are hard by nature, as our virtuosi do between petrified shells, & those which are lapides sui generis"?[11] Or Jane Austen's too-rational Sir Edward Denham, of the unfinished *Sanditon* of 1817, who, in a characteristic tirade against novels ("those puerile Emanations"), complained that "in vain may we put them into a literary Alembic;—we distil nothing which can add to Science," although Austen and her readers would have shunned the implication.[12] Novels, after all, offer the finest calibrations with which to measure the fluctuations of socioeconomic pressure.

In a more sympathetic tone, meanwhile, Sarah Hoare's hymn to a sea conch was awash with harmoniously specialist terms:

> Gracefully striate is thy shell,
> Transverse and longitudinal,
> And delicately fair[13]

while Goethe based the structure of an entire novel on the metaphor of chemical attraction. His Eduard and Charlotte, the romantic catalysts of *Elective Affinities* (1809), bond helplessly to one another like a pair of affiliated molecules. They are not themselves so much as the agents of an irresistible natural force. Austen's Sir Edward Denham, no doubt, would have considered their escapades unutterably foolish.

This world of the natural sciences was an open mine of similes, quarried to supply the increasingly expressive requirements of the age. Elizabeth Montagu's heart, like Alexander Pope's grotto at Twickenham, was encrusted with a geology of sentiment. Little wonder that Coleridge attended Royal Institution lectures in order, as he put it, "to renew my stock of metaphors."[14] He himself was to offer new words to the language, coining the term *psychosomatic* after watching his hero, Humphry Davy, cure a case of suspected paralysis by administering a course of placebos.[15]

This was what the future must have looked like for the many who crowded the auditoriums of the burgeoning theater of science: not just a parade of man's ever increasing familiarity with the mutable territories of nature, but the unalloyed joy of their discovery and naming. Here was a cultural scene that delighted in both the unraveling of the processes of nature and the languages forged in the attempt. As new forms of understanding emerged, new forms of expression, both literal and metaphorical, appeared alongside them to support them in their work. The new ways of seeing became increasingly bound up with the new formulations of words.

This was the climate, with its as yet unchanged belief in the positive virtues of science and scientific thought, into which Luke Howard was due to release his classified names for the clouds. And his language, too, the language of the skies, would rise to enrich the wider cultural climate of the age. It would give weight to the weightless forms of the air, institute a transformation of outlook and expression, and alter forever the relationship between the world and the restless, overarching sky.

Chapter Two

A Brief History of Clouds

They are the celestial Clouds, the patron goddesses of the layabout. From them come our intelligence, our dialectic and our reason.

Aristophanes, c. 420 B.C.[1]

Onlookers had, of course, attempted to account for the appearance and feel of their weather since the remotest era of antiquity. Weather has always been the main determining aspect of man's environmental experience, with its ambiguities forever demanding renewed interpretation and debate. The sky throughout history has been variously filled by the promptings of the imagination, whether with gods and prophecies and the rhythms of the zodiac, or with the first faint stirrings of scientific thought.

Certainly, some of the world's earliest written documents record attempts to come to terms with the endless variabilities of weather. Ancient Egyptian, Chaldean, and Babylonian texts, preserved on fragile squares of papyrus and clay, speak of the mysteries of clouds, thunder, and rainmaking and include the earliest fragmented attempts at forecasting and forewarning. "When a dark halo surrounds the moon, the month will bring rain or gather clouds," declares a four-thousand-year-old Chaldean prophecy. "When a cloud grows dark in heaven, a wind will blow," forewarns another.[2] These texts and others like them, the earliest weather forecasts on record, may well have been pieces of already ancient meteorological lore, but there is no way of knowing whether they also referred to imminent changes in the broader social climate of the times. All we can read in them now is the air of palpable misgiving.

Farther east, meanwhile, during the Shang dynasty, Chinese scholars were keeping more descriptive weather journals and attempting to analyze the contents in ten-day batches, of which a few faded traces have survived. In them, sightings of rainbows, halos, and mock suns were recorded while the direction of the prevailing wind was logged. Levels of rainfall and snowfall were measured, too, the latter by bamboo snow gauges sited high in the mountains of the northern provinces, "the Lame Dragon's frozen peaks," in the words of a poem of the times,

> Where trees and grasses dare not grow;
> Where the river runs too wide to cross

And too deep to plumb,
And the sky is deep with snow.[3]

Ingenious prototype hygrometers were also developed in the hills of ancient China, which used the absorbent qualities of charcoal: its dry weight was taken and then compared with its damp weight after timed exposure to the air. Humidity levels were then calculated from the difference. Here, more than two thousand years before the advent of the Christian era, some of the emerging certainties of the natural world were beginning to be tested and logged.[4]

All scientific advances are part of wider social developments, and what lay behind the sophistication of early Chinese meteorology was the nature of the evolving Chinese worldview. The developing doctrine of yin and yang, the twin fundamental forces that serve to balance everything in creation, was becoming an increasingly important component of Chinese natural and moral philosophy. By the end of the fourth century B.C., yin had become associated figuratively with clouds and rain (as earthbound elements of the female principle), while yang had become associated, equally figuratively, with fire and the heat of the sun (as celestial elements of the balancing male principle). According to the doctrine, these complementary properties were never to be found separately in nature, although one tended to dominate the other at any given time.

All change on earth, all terrestrial transformations and patterns made by nature, including, of course, the weather, could be understood through these emerging foreground ideas, and Chinese meteorology (along with its neighboring disciplines of astronomy and mathematics) developed partly as a means to lend expression and reinforcement to this sublimely harmonious philosophy.

The symmetry of the water cycle, for instance, offered a perfect working analogy of the doctrine of elemental change and cooperation: the warmth of the (yang) sun nourishes the (yin) clouds through the semisecret agency of evaporation. The ceaseless rise and fall of water through evaporation, condensation, and precipitation echoes the balance of harmony and exchange that underlay the entire working structure of the Chinese mental universe. Even the violence of electrical storms served to illustrate this leveling out of all forms of natural energy: an overabundance of (yin) rain requires a compensating bolt of (yang) fire to balance things in the teeming atmosphere. Hence, the rain cloud's extravagant gifts to earth of thunder, lightning, and highly charged traces of sulfur. They come as the visible payment of an energy debt built up over time, high in the farthest reaches of the air.

A few centuries later the Taoist religion developed an entire Ministry of Thunder for its pantheon. This divine administration included the gods of Thunder and Lightning, the Earl of Wind, and the Master of Rain and his young apprentice, Yun-t'ung, the Little Boy of the Clouds, whose job was to keep the floating reservoirs piled up, replenished, and full. Modern ideas of feng shui ("wind and water") living come from the long shadows cast into our own times by the undiminishing strength of these ideas.

In sharp contrast to such harmonious conceptions, much of the moral power of the Old Testament religion derived instead from the narration of extreme and persecutory weather events. From the Flood of Genesis to the plagues and hailstorms of Exodus, the books of Moses and the Prophets resound with the terrible weather of vengeance, much of which was brought to pass by violent winds from the east. Climate, the most traumatic of these episodes seem to say, is the one great precondition on earth, the one persistent feature of the natural world that cannot, will not, be controlled. Pestilence and predestiny alike descended from the darkening and omen-haunted skies, brought on by unfavorable disturbances in the jet stream.

The journeys of the children of Israel were prompted by the appearances of God in the clouds: over the Red Sea, over Mount Sinai, and, in their most dramatic manifestation, in anger above the tabernacle following the worship of the Golden Calf. For the troubled exiles of Israel, the great wrath-filled thunderheads that seemed to attend their every move symbolized not only the conditions attached to the renewal of their covenant with God but the shock of their uprooting from

Egypt. Within weeks of the outset of their journey, they had encountered geophysical extremes of famine and drought, followed by intense, heavy cloudbursts from on high.

Coming as they did from a lowland culture that relied on irrigation from passing rivers rather than from the endlessly stable blue skies above, such uncertainty was a forbidding glimpse into the shape of their future life. With an average annual rainfall in Egypt of a mere 25 to 50 millimeters, the phenomenon of the cloudburst had been relatively unknown, and having lived on its plains for more than four hundred years, the departing Israelites never would have encountered major cloud structures if their lives had remained undisrupted by flight. The take-up of water from the low-lying alluvial areas was not great enough, and the air above them was too hot and too clean to condense locally produced vapor into clouds, while rain-bearing systems traveling up along the trade winds would have exhausted themselves long before reaching the deltaic lands of the pharaohs.

The exodus to the semiarid zones of Sinai and Canaan, however, brought them into contact not only with the seasonal uncertainties of rainfall (with an annual average of more than 400 millimeters) but also with the unfamiliar sight of high-built convective cumulonimbus structures, the divine "pillars of cloud" that appeared to them as soon as they had left the deltaic lowlands of Goshen. The impact was sudden, and it hardly diminished over time as the clouds continued to direct their steps toward their long-awaited, long-forsaken home.[5]

The spectacle of clouds, in all their terrible unpredictability, came to symbolize the uncertainty, the strangeness, and the promise of their homecoming for the nomadic authors of the books of Moses. To learn the secrets of rainfall irrigation, as those lowland farmers came to realize, they would have to learn the wider secrets of the skies. And we can hear their questions, asked with mounting urgency in the books of Enoch and Job ("Hath the rain a father?"; "Who can number the clouds in wisdom?"; "Can any understand the spreadings of the clouds?"), reverberating down the centuries, dodged or unanswered or molded into a cosmology, in a great chain of reasoning that led on through a series of landmarks in meteorological history, one of the most important of which was the dramatic punctuation point of December 1802, when the clouds were first properly named by a 30-year-old chemist in a basement laboratory in London.

In the meantime, more rational attempts to penetrate the mysteries of the upper elements arose in the hope of bringing them nearer and thereby reducing their ancient, atavistic threat. It has not always been reassuring to regard the skies as the haunts of the vengeful gods, although most cultures have, at one time or another, configured the towering clouds as the homes of the gods of the lower and upper air. In dividing the world below from the world above, clouds have served the mythmakers well as the boundary markers of other realities. And what other worlds might not be hidden from our sight beyond the high threshold of the veiling clouds?

In Norse mythology, for instance, Frigg (the wife of Odin), whose name translates as "well beloved," was represented as the goddess who ruled from above the clouds. High up in Fensalir, her Hall of Mists, with wheel and distaff and infinite patience, she spun the golden threads that were woven by the winds into the bands of pink and orange cirrostratus, glimpsed at daybreak and at sunset the world over. These higher clouds were her special preserve, untouched by the hands of the other gods, while at the creation, the brains of the frost giant Ymir had been flung into the summer air to form the familiar low-lying cumulus clouds. All this was according to Alvis, the all-knowing dwarf, who recited the names by which "the clouds, that hold the rain, in each and every world" were known:

> Men call them Clouds; the gods say Chance of Showers and the Vanir say Wind Kites. The giants name them Hope of Rain, the elves Weather Might, and in Hel they're known as Helmets of Secrets.[6]

This early, summarized classification (in Kevin Crossley-Holland's luminous translation) reveals a taxonomy of weather observer broken down into matching cloud types (as well as, elsewhere in the passage, matching wind, star, and sky types). Just as the fair-weather cumulus suited the sunny outlook of the Vanir (the fertility gods who invented magic), brooding sheets of stratocumulus mantled the ghastly citadel of Hel. Clouds, as the mythmakers of all cultures have discovered, make the perfect backdrops to their emblematic images of cosmogony. Neutral, varied, and aloof, they crawl with a potentially unending array of autogenetic divinities.

But the ancient Greek philosophers, much disposed toward looking up at the sky for answers, were keen to account for the variety of its moods and appearances without having to make an open appeal to divinity. Their philosophical project saw the universe itself as a series of physical problems to be pondered at a separate remove from morality. Among the various fields that arose in the pre-Socratic sixth and seventh centuries B.C. were astronomy (the study of celestial bodies), brontology (the study of thunder), ceraunics (the study of lightning), and nephology (the study of clouds). All were aspects of a connected scheme of study that, with its many beginnings and endings and its neighboring distractions of astrology and prophecy, has proved to be one of the longest-lived and most varied research projects in the history of natural inquiry.

The first major phase of Western nephology (in its written form, at least) began with Thales of Miletus, one of the seven great sages of pre-Socratic thought, and ended in the middle of the seventeenth century, after René Descartes had prized physics away from the deadening grip of Aristotle. After this date it developed sporadically, in the wake of the uncoverings of the central laws of physics. Howard himself grew interested in tracing this hidden history of nephology, and what follows would largely have been known to him, albeit much later in his career, when reading and correspondence had taken the place of original meteorological research. Howard was to find it an enthralling subject, spending much of his semi-retirement in his study at home, where he lined the walls with his growing collection of ancient books about the weather.

It was there that he read about Thales of Miletus (c. 625–545 B.C.), a figure widely regarded as the first real "scientist" to have been produced by Western civilization. Thales was interested principally in mathematics and astronomy and was famed for his successful prediction of a solar eclipse in 585 B.C., but he was also an impressive meteorological theorist. Like the earlier Chinese (whose ideas may well have traveled west along the trade routes to Grecian Ionia), he entertained a semi-mystical reverence for water as the giver and sustainer of life on earth. This reverence, combined with the Homeric belief that the earth was held afloat by a universal aqueous bed, led him to a mental map of a world based entirely on water, nourished and defined by water's life-giving properties. His thoughts on the mobility of this "material principle," as it rose and fell between heaven and earth, constituted in essence an early and accurate description of the water cycle, although it seems unlikely that Thales could have known anything of the principles of cooling, condensation, or cloud formation.

But in maintaining that everything in nature was to a greater or lesser degree a modification of water, Thales had voiced a fundamental truth about human existence: that we live not, in reality, on the summit of a solid earth but at the bottom of an ocean of air. Even though he maintained that all this air was a form of another element, water (water vapor in fact accounts for less than 2 percent of our atmosphere), the idea of studying the earth's surrounding film of gases was the first real step toward the formation of the meteorological, as distinct from the astrographical, imagination. Meteorology as a distinct and recognizable branch of knowledge, which deals with the atmosphere and all it contains, had been formally introduced to the Western mind.

Thales' fame and scientific influence soon attracted followers, prominent among whom was his fellow Milesian Anaximander (c. 610–547 B.C.). Anaximander was the author of one of the earliest scientific treatises in history, in which he was the first to suggest that lightning was caused by friction within the bodies of clouds, as well as the first to describe wind as "a flowing of air."[7] Although his descriptions of weather were subordinated to his philosophical notions of the infinite and indefinite origins of matter, they were, in their own right, perceptive accounts of specific events in nature. His idea that pockets of air can be pushed apart, gathered together, and violently moved about by unnamed forces was a fair summary of the means by which wind is generated: from the local upward and downward convections of air over landforms (following the rise and fall of ground temperature) to the huge onrushes made as it streams from areas of high pressure into areas of low pressure, like air escaping from a tire. Francis Beaufort was later to be inspired by both Anaximander and Luke Howard when he came to frame his grading of the force of the wind, in a successful amalgamation of related ideas that were separated by two thousand years. But in a typical twist of fate, one shared by many ancient hypotheses, Anaximander's perceptive descriptions had to wait out those twenty centuries for acceptance.

The Greeks had no meteorological instruments with which to confirm or refute their observations of nature, but in a sense this hardly mattered, as their genius lay more in the questions they asked than in any of the answers they hazarded. Yet they did recognize a practical as well as a theoretical dimension to meteorological knowledge, and by the fifth century B.C. weather bulletins were displayed on columns in the public squares of many of the cities of the Mediterranean. These "peg" almanacs (called *parapegmata*, from the Greek *parapegma*—"to fix into") formed eyewitness records of local conditions, with the occasional attempt at forecasting thrown in. Typical examples, such as "the shoulders of Virgo are rising," "Rising of Arcturus; south wind, rain and thunder," and "the weather will likely change," confirm the astrological context of the majority of ancient forecasts: they also confirm that the maddening caution of the weather forecaster, who will blithely, and in all seriousness, offer a 24-hour picture that includes the entire spectrum of possible weather, all the way from heat wave to hail, has been with us since early antiquity.[8]

Information from such sources as the peg almanacs furthered the growth of comparative science, which arose as travelers returned from their wanderings with new observations to impart. Democritus (c. 460–370 B.C.), for example, one of the Western world's first great travelers and observers, developed a theory of cloud formation in tandem with his atomistic picture of the universe. He described a world consisting of an infinite number of tiny, inert particles moving in random mechanistic patterns, and conceived his theory of clouds after witnessing the melting snows of the summer solstice during his travels in northern Europe. The vapors from the meltwater, he suggested, after giving the matter some thought, were raised aloft and then carried south by the Etesian winds, from where they returned to earth as the rain that watered (and regularly flooded) the lakes feeding into the Nile. This was not merely an elegant rehearsal of the image of the endless water cycle; it was also an early and prescient description of the movement of storm systems around the world. It was, however, yet another idea like Anaximander's thoughts on wind that didn't reemerge in any serious way until twenty centuries had passed.

It is hard to say how much of this neglect can be put down to the simple misfortune of having conceived of thoughts far ahead of their time, or how much can be credited to the overwhelming influence of a new synthesis of thought that emerged in the middle of the fourth century B.C. from the work of Plato's most famous and prolific pupil: Aristotle (384–322 B.C.).

Although only about thirty or so of Aristotle's treatises have survived, collectively they express a single, outstanding conviction that would have run like a thread through the lost other works: the conviction that everything in nature has ultimate order and intelligibility and is tending toward its natural state. This was true even of those things in the world that persist in a state of change. And in the *Meteorologica*, a treatise he composed around 340 B.C., Aristotle addressed himself specifically to the most comprehensively changeable, and thereby the most challenging, of any of nature's aspects.

Unlike earlier Chinese thought, which viewed all natural processes in terms of transformative exchanges, fourth-century Greek philosophy had come to regard the notion of change as an affront to the harmonious ordering of the universe. Plato, for example, held that the changeability of matter was merely a symptom of its material imperfection. "Forms" themselves (these were not individual objects, but rather their incorporeal cosmic ideas) remained perfect and therefore, by extension, remained changeless. Real knowledge, real science, sought to apprehend only this autonomous realm of forms. Platonic knowledge did not deal in particulars, and to seek to understand lowly, material events such as the wind or the rain or the movement of the clouds was held to be something of an indulgence. Like Johnson's account of an English conversation, it was nothing more than talking about the weather. And such talk did little to support the idea of the intrinsic static order of the cosmos.

Aristotle, however, departed radically from these Platonic ideas (as well as from the earlier atomism of Democritus) to claim instead that everything in the sublunary world was, is, and forever would remain in a permanent state of flux.[9] And the jostling clouds would prove to be one of Aristotle's foremost examples, brought in to uphold the truth of his philosophical departure.

To bear out the conviction that nature was embedded in essential change, he effected a root-and-branch reorganization of its principal means of inquiry: biology, physics, and meteorology were reordered and refined along newly drawn lines of thought. In meteorology, as in the other sciences, Aristotle sought to emphasize the role played by the four ancient elements of earth, air, fire, and water and their associated, paired qualities of heat and cold, dryness and moisture, which were "exhaled," as he conceived it, from the physical materials of the earth. The warm and moist exhalations, for example, are responsible for the production of clouds and rain, while the dry exhalations are responsible for driving thunder and wind around the earth. All manifestations of natural change then take place between these pairs of elemental opposites through which all conditions in the world are determined. Each element, with earth at the center of the spherical arrangement, has (or actively seeks) a rightful place in the natural world. Buoyant air, for example, seeks to rest in the sphere above water but below the sphere of fire, while water moves and eddies as it searches for its place above the earth.

To show the full explanatory force of these ideas, Aristotle began the treatise on meteorology with an introductory case, the formation and life span of clouds:

> Let us first deal with air, and approach the solution of our main problem by means of a discussion of the question, why do not clouds form in the upper air as one might on the face of it expect?[10]

The answer, according to Aristotle, depended on the mingling of the stratified elements: the heat of the sun rearranges the cold water on the surface of the earth to form a new, warm substance, similar to air, which then rises through it. This is the material from which clouds are formed. But, went the theory, the layers of air above the highest mountains contain too much fire for clouds to be formed there; similarly, the layers of air found close to the ground also contain too much reflected sunlight for clouds to form. Clouds only gather "where the rays begin to lose their force by dispersion in the void."[11] In other words, there is a distinct cloud-forming layer to be found about halfway between the surface of the earth and the upper atmosphere, in which the balance of the elements of air and water and their properties of moisture and warmth finds ideal resolution in favor of the production of clouds.[12]

As a mechanistic explanation based entirely on physical assumptions, it could hardly have been bettered, and this was true of the other explanations proffered throughout the course of the book. These, and the immense authority established by Aristotle in other fields of thought, ensured the continued success of his treatise. It went on to dominate Western meteorological thinking for the following two thousand years, although his influence began to wane during the early Christian era, when ancient Greek ideas sank almost entirely from view.

But long before then, Roman authors such as Seneca, Pliny, and Lucretius had sought to preserve the scientific spirit of the Greeks by compiling anthologies of natural knowledge from both new and surviving sources. Seneca, in his *Natural Questions*, a ten-book digest begun around A.D. 62, was much impressed by the dynamic energy exhibited by the weather, and viewed clouds as an integral part of atmospheric changeability:

> It is always flowing and its periods of rest are short. Within a brief moment it changes into a condition other than the one in which it had been; now rainy, now clear, now a varied mixture between the two. Clouds—which are closely associated with atmosphere, into which

atmosphere congeals and from which it is dissolved—sometimes gather, sometimes disperse, and sometimes remain motionless.[13]

Clouds, for Seneca, were the very engines of weather, responsible for thunder, lightning, and the phenomenon of the rainbow. The atmosphere itself was both built and dismantled by the actions of the opalescent clouds. But still no explanation other than Aristotle's exhalations was available to account for their initial formation.

Pliny, too, who lost his life in the great eruption of Vesuvius of A.D. 79, summarized clouds as a blend of materials drawn from both the upper element of air and "the unlimited quantity of terrestial vapor" that girdled the earth before making its way into the atmosphere. He summarized the entire meteorological life cycle as a great rising and falling, which occurred in response to the attractive and repulsive power of the stars: "rain falls, clouds rise, rivers dry up, hailstorms sweep down; rays scorch, and impinging from every side on the earth in the middle of the world, then are broken and recoil and carry with them the moisture they have drunk up. Steam falls from on high and again returns on high."[14] Nature swings to and fro like a vast pendulum, on a course that is frequently troubled by the velocity of the stars and the planets. Clouds ebb and flow in the wake of their unpredictable passage. But again, all this Plinian turbulence was little more than an elegant restatement of Aristotelian change.

The Epicurean philosopher Lucretius, on the other hand, a man who reveled in the sensory joys of nature, made a more directly idiolectic attempt to account for the formation of clouds. A "sudden coalescence, in the upper reaches of the sky, of many flying atoms of relatively rough material, such that even a slight entanglement clasps them firmly together" was responsible for the small initial structures, which then banded together to form the larger sorts of clouds. These then grew by mutual fusion until dispersed by storms and rain.[15] This was the atomistic model of cloud formation, blended from the ideas of Democritus and Aristotle, that would be revived in the eighteenth century.

But as Christianity began to spread throughout the continent of Europe, the scientific influence of ancient authorities such as Aristotle, Pliny, and Lucretius declined. Atmospheric activity was returned to a context of divine and moral intervention. "The Lord hath his way in the whirlwind, and in the storm," says the first chapter of the Book of Nahum, "and the clouds are the dust of his feet." For centuries, the explanatory power of such pronouncements served to overrule any vestiges of ancient natural knowledge. It was not until medieval thought began to reconcile Christian theology with earlier Aristotelianism that leading scholars such as Adelard of Bath or the Venerable Bede were in a position to compile tracts on the processes of nature. Bede's *De Natura Rerum*, composed in Northumbria sometime between the 690s and the 730s, was itself largely a restatement of the Plinian restatement of Aristotle, filtered through the earlier natural theology of Isidore of Seville.

Such, then, was the state of meteorology in the long centuries after the decline of Greece and Rome: little or no original thought, merely an anthologizing deference to the authority of ancient sources.

But by the middle of the seventeenth century and with the coming of the scientific revolution, the Aristotelian view of the world at last found itself coming under serious threat from new ideas about the nature of the cosmos and the actions of the elements. Aristotle's hold on meteorology was brought to an end largely by the insights of one man: the French Jesuit philosopher René Descartes (1596–1650).

Descartes spent much of his life in search of peace and quiet and chose to live in Holland to achieve it. There, alongside his wider philosophical project of making himself master of first principles, he developed his long-held interest in meteorology, intended as a means to furnish examples in support of the rational, anti-Aristotelian approach to nature that he devised and outlined in his *Discourse on Method* in 1637. The *Discourse*, as it claimed from the outset, would demonstrate the way to certain knowledge. Certain clear and fundamental ideas would open out to an understanding of the phenomena of nature. To shore up this claim, an essay, "On Meteors," was appended to the first edition as a working model of the method, designed to show how it could be used to approach any natural phenomenon—even one as discontinuous as clouds.

As had Aristotle in the *Meteorologica*, Descartes took clouds to be a primary case in point, viewing them as testing grounds for new ideas in the philosophy of nature. If you can philosophize about the clouds, he felt, then surely you can philosophize about anything. Descartes found clouds to be especially interesting not so much for their physical properties, which for him were routinely mechanistic, but for the challenge that they offered to the senses. This included the sense of wonder: "We have naturally more admiration for things that are above us than for those that are at the same height or lower," he observed at the beginning of the treatise, in an acute analysis of the hierarchy of sight. But it was exactly this mystification of the clouds that he sought above all to reduce:

> Since one must turn his eyes toward heaven to look at them, we think of them as being so high that even poets and painters see them as the throne of God and pretend that He uses his own hands to open and close the doors to the winds, to sprinkle dew on the flowers, and to hurl lightning against the rocks. That makes me hope that if I explain their nature here so that one will no longer have occasion to admire anything about what is seen or descends from above, one will easily

> believe that it is possible in some manner to find the causes of everything wonderful above the earth.[16]

If the clouds could be rationally and convincingly explained, without recourse to superstition or prejudice, then so could anything else in nature, for they represented the most extreme manifestation of the ungraspable. They were, indeed, so difficult to decipher that they might lead one more generally to distrust the judgment of one's senses, so easily misguided by changeable objects in the atmosphere.

Clouds, in short, were aids to philosophy, and Descartes painted a convincing picture of their operation on behalf of his method. His clouds, combative yet strangely dissolute, sidled like thoughts into the minds of their observers.

Descartes went on to consider the physical nature of clouds, which, he argued, most likely consisted of water droplets or small particles of ice formed by compressed vapors given off by objects on the ground, rather than by Aristotle's mingled "exhalations." These droplets or particles then grow through coalescence, "which as soon as they are *joyned*, rise up in little heaps, and these gather'd together compose vast Bulks . . . so loose and spungy, that they cannot by their weight overcome the Resistance of the Air."[17] When the drops have grown too large to stay aloft, they fall as rain (when the air temperature is warm enough), or as snow or hail (when the air temperature is too cold for rain). Descartes had deduced that clouds were always cold enough for snow ("a *Cloud* is nothing else, but a great heap of *Snow* close clinging together") and believed that hail was formed from ice particles that had melted and then refrozen in cold air on the way down.[18] In common with many of his mechanistic descriptions of the workings of nature, Descartes's theories of clouds and rain, as well as his theories of lightning and the rainbow, were both original and magnificently expressed.

As Aristotle's exhalations were finally breathing their last under the weight of Cartesian deduction, new technologies were moving in to hasten their end. The span of Descartes's life had seen the six instruments introduced that would determine the future direction of all scientific investigation. After the appearance of the telescope, the microscope, the air pump, the pendulum clock, the thermometer, and the barometer, all in the first half of the seventeenth century, scientific inquiry would never be the same again. Shared methodologies, whether in the field, the laboratory, or the private museum, would arise as the means to apprehend the teeming world of things. Meteorology shared in this gathering sense of advance, and, in concert with the rise of other kinds of measurers and compilers, the era of the weather collector had begun.

Yet bafflement over clouds and cloud formation proliferated during the post-Cartesian era of the scientific revolution just as freely as it did among the pre-Socratic philosophers. The full range of scientific outlooks, from unreconstructed

Aristotelianism to alchemy to modern Newtonian physics, embraced nephological theorizing, much of which was remarkable only for its internal systems of logic. The "menstruum theory," for example, named after the alchemical term for a solvent, held that acids in the air were responsible for corroding water into clouds and keeping them suspended, just as they corroded and clouded the surfaces of metals. Another theory maintained that particles of fire became detached from sunbeams and adhered to particles of water in order to manufacture special, lighter-than-air molecules that rise to join up as clouds; rain is then produced by the separation of the fire particles, which releases the water from the cloud to fall under the influence of gravity.

Oliver Goldsmith, on the other hand, in his *History of the Earth, and Animated Nature*, compiled in the late 1760s, held that clouds were the product of a repelling power in nature, with vapor hurled into the atmosphere by the burning rays of the sun, like steam spitting up from a hot iron plunged into water. The clouds then grow like rolling balls of vapor until broken up into rain by resistance from the wind. He seemed to have little confidence in his explanation, however, as he later conceded that "it would be to very little purpose to attempt explaining how these wonders are effected."[19]

Still other hypotheses likened vapor in the air to "a lump of sugar dissolved in a cup of tea," the power of the wind acting to stir it into solution.[20]

Goethe, though, believed that the earth's gravitational field was responsible for bending atmospheric water vapor into its variety of shapes and forms, and, as we will see, he abandoned this view only when he came to read a translation of Luke Howard's essay in 1815.

The most widespread theory of all, however, and the one against which Howard was to do continual battle, was the vesicular, or "bubble," theory of clouds. This theory maintained that particles of water, through the action of the sun, form into hollow spherules filled with an "aura," or a highly rarefied form of air, which, becoming lighter than normal atmospheric air, rises like an air balloon to form itself into clouds. Rain begins to fall when these bubbles finally burst. This would have been the view held by the majority of the Plough Court audience, and it was the view that Howard was determined to explode.

All these rival explanations had their problems, as even their proponents were aware. Oliver Goldsmith wryly conceded that "every cloud that moves, and every shower that falls, serves to mortify the philosopher's pride, and to show him hidden qualities in air and water, that he finds difficult to explain."[21] And this was true; yet it was just such an explanation that Luke Howard furnished on that incandescent evening in the London laboratory, where the history of nephology—the history of clouds—was about to be dramatically rewritten.

Chapter Three

The Cloud Messenger

The fairest things are those which live,
And vanish ere their name we give;
The rosiest clouds in evening's sky,
Are those which soonest fade and fly.

Mary Russell Mitford, 1811[1]

"In tracing the various appearances of clouds, I have only adverted to their connection with the different states of the atmosphere, on which, indeed, their diversity in a great measure depends. I have purposely avoided mixing in difficult and doubtful explanations with what is only, after all, a simple descriptive arrangement. There is much more to be done in the field, as I have sought to intimate already, but these, at least, are the clouds as I know them. Or, perhaps I should say, these are the clouds as I have so far understood them. I thank you for your attention and I hope that, as usual, there might be plenty of time for discussion."

Luke Howard had been speaking for nearly an hour, during which time his audience had found itself in a state of gradually mounting excitement. By the time he reached the concluding words of his address, the Plough Court laboratory was in an uproar. Everyone in the audience had recognized the importance of what they had just heard, and all were in a mood to have it confirmed aloud by their friends and neighbors in the room. Over the course of the past hour they had been introduced not only to new explanations of the formation and life span of clouds, but also to a poetic new terminology: *Cirrus, Stratus, Cumulus, Nimbus,* and the other names, too, the names of intermediate compounds and modified forms, whose differences were based on altitude, air temperature, and the shaping powers of upward radiation. There was much that needed to be taken on board.

Clouds, as everyone in the room would have known already, were staging posts in the rise and fall of water as it made its way on its endless compensating journeys between the earth and the fruitful sky. Yet the nature of the means of their exact construction remained a mystery to most observers, who, on the whole, were still in thrall to the vesicular, or "bubble," theory that had dominated meteorological thinking for the better part of a century. The earlier speculations, in all their strangeness, had mostly been forgotten or were treated as historical curiosities to be glanced at, derided, and abandoned. Howard, however, following Descartes, was adamant that clouds were formed from actual solid drops of water and ice, condensed from their vaporous forms by the fall in temperature they encountered as they ascended through the rapidly cooling lower atmosphere. Balloon pioneers during the 1780s had confirmed just how cold it could get up in the realm of the clouds: the temperature fell some 6.5° C for every thousand meters they ascended. By the time the middle of a major cumulus cloud had been reached, the temperature would have dropped to below freezing, while the oxygen concentration of the air would be starting to thin quite dangerously. That was what the balloonists meant by "dizzy heights."

Howard was not, of course, the first to insist that clouds were best understood as entities with physical properties of their own, obeying the same essential laws that governed the rest of the natural world (with one or two interesting anomalies: water, after all, is a very strange material). It had long been accepted by many of the more scientifically minded that clouds, despite their distance and their seeming intangibility, should be studied and apprehended like any other objects in creation. Howard's view thus seemed, on the face of it, broadly Cartesian in reason and scope.

There was more, however, and better. Luke Howard also claimed that there was a fixed and constant number of basic cloud types, and this number was not, as the audience might have expected, in the hundreds or the thousands, like the teeming clouds themselves, with each as individual as a thumbprint. Had this been the case, it would have rendered them both unclassifiable and unaccountable—just so many stains upon the sky. Howard's claim, on the contrary, was that there were only three basic families of cloud, into which every one of the thousands of ambiguous forms could be categorized with certainty. The clouds obeyed a system, and once recognized in outline, their basic forms would be "as distinguishable from each other as a tree from a hill, or the latter from a lake," for each displayed the simplest possible visual characteristics.[2]

The names Howard devised for them were designed to convey a descriptive sense of each cloud type's outward characteristics, a practice derived from the usual procedures of natural history classification, and were taken from the Latin, for ease of adoption "by the learned of different nations": *cirrus* (from the Latin for "fiber," or "hair"), *cumulus* (from the Latin for "heap," or "pile"), and *stratus* (from the Latin for "layer," or "sheet"). Clouds were thus divided into tendrils, heaps, and layers: the three formations at the heart of their design.

This was the point at which some of the audience members sat upright in their chairs, as a murmur made its way around the room. It was clear that a brilliant lesson was unfolding and that not a word of it ought to be missed.

Howard went on to name four other cloud types, all of which were either modifications or aggregates of the three major families of formation. Clouds continually unite, pass into one another, and disperse, but always in recognizable stages. The rain cloud *nimbus*, for example (from the Latin for "cloud"), was, according to Howard, a rainy combination of all three types (although *nimbus* was reclassified as "nimbostratus" by meteorologists in 1932, by which time the science of rain had developed beyond all recognition).

The modification of clouds was a major new idea, and what struck the audience most vividly about it was its elegant and powerful fittingness. All that they had just heard seemed so clear and so self-evident. Some must have wondered how it was that no one—not even in antiquity—had named or graded the clouds before, or, if they had, why their efforts had left no trace in the language. How could it be that the task had been waiting for Howard, who had succeeded in wringing a kind of exactitude from out of the vaporous clouds. Their forms, though shapeless and unresolved, had at last, it seemed, been securely grasped. Howard had given a set of names to a radical fluidity and impermanence that seemed every bit as magical, to that first audience, as the Eskimos' fabled vocabulary of snow.

Everyone in the room was struck by the realization of their presence at a moment of clarity, a moment when the world seemed suddenly to have been pulled more sharply, more richly, into focus. It was a moment of unparalleled insight: in naming the clouds, in bestowing their language over the course of an evening's lecture, Luke Howard had presided over a defining event in a new kind of natural history. Modern meteorology, hitherto a slower developer than many of its neighbors in the natural sciences, had been given a means to spring into new scientific shape. Howard's insight had opened up the clouds to view so they could now be seen for what they were: the visible signs of the otherwise hidden movements of the atmosphere.

Nineteenth-century meteorology took off with a public conversation set high amid the region of the clouds. It was a bold beginning both for a new century and for a new science.

And so the young Luke Howard soon found himself surrounded by his friends and his hosts, and by a pressing assembly of strangers. Over fifty people were in the room, which was warm now in spite of the season, and noises were becoming hard to distinguish. Chairs were scraped, voices were raised, people turned to address their friends and acquaintances, or called out questions to the distracted young lecturer, to this, what was it, this *nephologist* they had listened to with growing admiration.

Not only his words but his drawings too had been well received, flourished as they were at the start of each new explication. There had been seven large sketches, done in pencil washed over with watercolor, each more impressive than the last, each illustrating a single cloud type suspended over varying landscapes or supervised by an overarching sky. These drawings, over which so much care had been taken, were the visual prompt for each new visionary account, fixing the word-hoard in the minds of his listeners. Here was cirrus, for instance, flying high, hanging serenely just as Howard had described it, "pencilled, as it were, onto the sky."

Cirrus, in the engraved version of Luke Howard's original drawing

The audience was susceptible to the twin persuasions of speech and spectacle, and the evening had provided both in great abundance. The clear light of a new insight had shone out into the evening air, and all in the room had been impressed by its flickering promise.

One man in particular, though, had been especially impressed.

Alexander Tilloch, a Scottish-born publisher and magazine proprietor, was a man whom Howard at earlier meetings had found a somewhat intimidating figure. He was a worldly character who used his connections as freely as he did his frequent bouts of sarcasm, employing both to equal effect. Now he used his height to press his way through to the hesitant new star, who he thought might well be interested in what he had to say. Tilloch grasped the young speaker's hand and boomed at him with unmistakable, if daunting, approval: "That's more like it, my boy, well done; why don't you come and see me tomorrow? I think the magazine can find some room for your clouds in a coming edition. Tomorrow it is, then." Tilloch, smiling broadly to himself, turned toward the door without waiting to hear the reply; but then Howard could not at that moment have supplied one. The warmth of the reception was distracting enough, yet now, he assumed, he had just been offered something more substantial. He had been offered the opportunity to work not merely on ideas for an evening's lecture, but on a serious and permanent contribution to knowledge for the leading scientific publication of the day. To be asked to contribute to the *Philosophical Magazine*, the best-known science journal in Europe, was the crowning event of the night's proceedings. He was going to publish his lecture on clouds.

The promise held out by the Scotsman's words grew quietly in Luke Howard's mind amid the surrounding clamor of the room. The essay would be expanded beyond its current bounds as a lecture, while the audience for his ideas and for his new and poetic terms would grow from fifty to maybe fifty thousand, given all the libraries, translators, and corresponding societies then at work around the increasingly learned world. "The preaching of Sermons is Speaking to a few of Mankind," as Daniel Defoe once wrote, "but the printing of Books is Talking to the whole World."[3] And Defoe, who had worked as both a journalist and a spy, had known more than most about the value of words on the page. The language that Howard had brought into fruition in that room—the sublime new language of clouds—was scarcely an hour old, but already it was talking its way into the world.

But these thoughts were for the future, a future that would begin with the next day's meeting with Alexander Tilloch: his editor. Luke Howard smiled at the thought. Meanwhile there was the rest of this happy, if exhausting, evening to get through, with all the competing noise of its compliments and congratulations. And it was late before the last remaining members of the audience began to make their way toward the door, out through the archway onto Lombard Street, and into the cold of the vast and starlit night.

Chapter Four

Scenes from Childhood

"What were the skies like when you were young?"
The Orb, "Little Fluffy Clouds," 1991

Luke Howard had been drawn from childhood by what he called his penchant for the pursuit of meteorology, the one pursuit that gave color and shape to what might otherwise have been a long and relatively uneventful life of duty and Dissent. It was through the act of looking up at the cloud-written sky that he confronted the face of his destiny, just as the young Walter Raleigh had once gazed out upon the beckoning, life-transforming ocean. There are certain individuals, it seems, who are called by the vastness of the elements, and who return an answering look, upward or outward, into the distant views that spread themselves out before their gaze. Thus they are drawn, while young and restless and full of dreams, closer to their visions of a possible future.

Howard was not alone in having developed an early enthusiasm for the study of the canvas of nature, but unlike the great John Hunter (1728–93), for example—who complained in his later years that when he had been a boy he had "wanted to know all about the clouds and the grasses, and why the leaves changed colour in the autumn," but that these had been questions which "nobody knew or cared anything about"—Howard had been born at just the right time and in just the right place to further his natural curiosity. He emerged, fortuitously, into the very crucible of public science.

Luke Howard was born on November 28, 1772, in a crowded house on Red Cross Street, in the Cripplegate Ward of London. Luke's father, Robert Howard (1738–1812), was a manufacturer of wrought iron and tinware, whose workshop in Clerkenwell was a flourishing and profitable concern. At any one time he employed a dozen or more mechanics and artisans, and his children all received an early introduction to the realities of the manufacturing life.

Robert Howard was a complex, authoritarian figure, who demanded a mixture of obedience and independent achievement from his children. He missed the softening influence of his first wife, Susannah Smith, who had been "pretty but delicate," according to family history, and who had died of consumption in her early twenties. The couple had first met when "Sukey," as he called her, came into his shop on an errand, and the looks that passed between them had made an immediate mutual impression. They were married within a matter of months and were devoted to each other's happiness, but the idyll was ended by her tragic death a few short years later. Three young children, John, Robert, and Joseph, were left in the unpracticed care of her husband. The task though proved too much for him alone, and he was soon remarried, to a woman he had known for years: Elizabeth Leatham (1742–1816), the daughter of a leading Quaker family from Pontefract, in Yorkshire. Robert's older brother Thomas had married her sister, and the two families had become close and affectionate. So after Susannah's death it was to Elizabeth that Robert turned for the companionship and love that she was only too happy to give.

Luke was the first child to be born to this second alliance of his father's. He was joined by four others over the following seven years, William, Isaac, George, and Elizabeth, only one of whom (Isaac) died in infancy, an unusually high survival rate for the time. The Howards and the Leathams were of a generally sturdy disposition, although the typhus epidemics of the 1790s would carry off Luke's older half brothers Robert and Joseph, who, like their mother before them, died while only in their early twenties. Their father's anxiety during their illnesses and his grief after their deaths were poured out in letters to his surviving children that make difficult reading even now, for there is something tragic in his transformation from a stern authority figure, pious and demanding, to a heartbroken ruin, sitting "with his handkerchief over his face, the tears trickling down his cheeks most of the day."[1]

In January 1791 he wrote to Luke to inform him that his brother Robert was nearing the end: "Thou wilt have little conception of the situation to which he is now reduced, he now lies in the drawing room where a bed was put up yesterday morning, the physicians desiring that he might be moved to the largest room, two windows, the door, and a window on the staircase all open, they don't care how much air, as still as possible."[2] The doctors came the next day, and the day after that, but could offer little hope of recovery. "My Dear Son," wrote Robert the elder to Luke on February 4, "this is a close visitation, afflicting as it is, I hope it will tend to our profit . . . what a proof is here of the preciousness of Health and of Time and of the advantages of an early dedication of Heart. I beseech thee make not light of this warning."[3] The warning, for Robert, had come too late. On February 8 he grew steadily worse, although "favoured with many lucid intervals." At one point he turned to his father and asked: "Father is Death upon me?" At nine in the evening he was heard repeating the Lord's Prayer, and at three the following morning "he quietly departed this Life—aged 24½ years." His parents were inconsolable, with Robert senior managing only a whispered "Farewell my dear child."[4] This time, there was nothing more to be said.

Robert Howard the elder had been born in Folkestone in 1738 and moved to London at around fourteen years of age to be apprenticed to a brazier in the City. The work was hot and physically demanding, but the fast-expanding market for ironware and tinware made it a profitable growth area, and one in which a bright young man might do well. As a child, Robert had received no formal education, but he was clever and motivated, with a naturally astute business mind. He also developed what was to prove, after his village-school education, a largely undernourished appetite for general knowledge. He sought to encourage this appetite in his children, too, although for him knowledge meant practical rather than pleasurable learning. As a Quaker, a small industrialist, and a prominent member of the Ironmongers' Company, he had little patience for the esoteric or the impractical. He would indulge his fourth son Luke's tastes for languages and the outdoor sciences on the understanding that his technical education would also be maintained. "Dost thou bend thy mind attentively to business," he counseled his son in a letter written in April 1790. "The diligent hand maketh Rich, and that not merely in increase but in

Peace of mind. It is a valuable thing on asking oneself, have I done my duty? to be able to answer Yes."[5]

Robert Howard was writing from experience. By the time that he had started to write advice-laden letters to his teenage sons, the tinsmithing and ornamental japanning business he had established in the 1770s had moved into the mass production of the Argand lamp, the precursor of the Victorian oil lamp. The Argand lamp, the brightest-burning yet cheapest model so far devised, was set to revolutionize the field of domestic lighting. By combining efficiency with affordability, it served to cast new light into the hitherto gloomy, candlelit interiors of the majority of European homes. The stench of burning tallow, a cheap candle wax rendered from mutton fat, was soon to be a thing of the past. With their affordable lamps installed in houses up and down the land, the Howards were already proving to be agents of the Enlightenment.

The commercial success of the Argand in Britain was to be the making of the Howard family fortune, although, as fourth-generation Quakers, they were forbidden by the discipline of the sect to enjoy most of the material comforts that might have flowed into the household. The wealth from the lamps was channeled into worthy causes, supporting the establishment of a Quaker school in Yorkshire and the work of the British and Foreign Bible Society, a charity cofounded by Howard senior in 1804. Theirs was a household in which luxury and idleness were held to be entirely against God's wishes. "What does idleness produce but mischief of every kind?" Howard senior was wont to remind his sons in further bulletins by letter. "Don't be afraid of business, thoult not be harder pressed than I have been; I have been long longing after leisure and don't yet see whether I shall ever overtake it."[6] So runs the lament of the hard-pressed father writing to his sons throughout history. Be more like me when you are young, he always wants to say, that you might live less like me when you are old. Like the majority of cross-generational advice, it is fated to be roundly ignored, and however young Luke might have agreed with the prescription, it was one that he would never find hard to overlook. His lifelong passion, as his own grandchildren were so fondly to recall, was for staring out of the window at the sky.

Yet passive as it might have seemed to his work-assailing father, this was the passion that was to bring about the dawn of the insight which was to make him famous. For it was while at school in Burford, near Oxford, at a stricter than usual Quaker establishment run by the formidable Thomas Huntley, that a set of everyday circumstances, unexceptional in themselves, coincided with a series of unusual events. The combined influence of these would serve to shape Luke Howard's nascent interest in the atmosphere, and in so doing would determine the scope of his future achievements in science.

Luke Howard's school life, up to this crucial point of change, had so far drifted by in a welter of unremarkable days. Pupils were granted only a four-week summer holiday, with the rest of the year spent living in the school itself. The schoolmaster, Thomas Huntley, was a family acquaintance who was gaining quite a reputation as an educator. The Hillside Academy was his own personal undertaking and was in considerable demand among the parents of the Dissenting communities of southern England, who regarded it as a highly progressive institution. In a sense they were right, although the pupils were less than enthralled by Huntley's irregular methods, which consisted, according to Howard, of beating those who could not learn fast enough, while leaving those who could to their own devices. Since Howard clearly fell into the second category, he was, as a consequence, left largely in control of arranging his own scheme of study.

Natural history took the foremost part, fed by fieldwork expeditions after hours. "The nutting season," as he recalled, "was a time of great delight; the long pinafore, then worn, was sewed up so as to form a bag, for the convenience of receiving the gathered nuts."[7] After a while Howard was joined at the school by two of his younger brothers, and they enjoyed going out to collect chestnuts together to roast over the schoolroom fire. He was happier outdoors than in, for indoors there was nothing he could do to avoid the hours of Latin grammar that stood at the heart of the compulsory curriculum. The declension of nouns went on for hours until he found he could do it in his sleep: *nubes,* a cloud; *nubis,* of a cloud; *nubi,* to or for a cloud; *nube,* by, with, or from a cloud, and so it went on. Yet it was the very repetitiveness of this routine that would lend such linguistic confidence later on, when he came to map out the Latin names for his new classification of clouds.

The main consolation for the boredom of his school days was afforded by the lodgings that he had been given, located at the back of the building. The view from high windows had already become one of his major indoor preoccupations, and the view from the windows of his hilltop school was particularly broad and extensive. Nearby fields and common land spread out beneath the wide Oxfordshire sky, as cloudscapes formed and unformed themselves before his watching eyes. "Already the bent of his own natural inclination was towards science," as an obituarist was to write of Howard's school days; "his observing eye had begun to be attracted by the varying beauties of the cloud-streaked sky, and something doubtless of the same pleasure was thus ministered to him, amid the lowland landscapes of Middlesex and Oxfordshire, which the child of the mountaineer derives from his daily friendship with the rocks and waterfalls of his home."[8] And it was only really while at school, stationed at the windows looking out over the view, that Howard began to pay his first real attention to the movements of the skies. He noticed how they shifted from moment to moment like the stirring eddies of a pool. Summer clouds reared up before his gaze in the distant space beyond the window glass, and as their shadows moved slowly across the fields, he felt drawn toward their sense of hurrying presence. It was as if they exerted upon his conscious mind a sort of pressure or inkling of recognition. In what might seem now to be a portent of his later achievements, he memorized the circumstances of "one remarkable configuration of the Clouds in a full sky, because it was of rare occurrence," and he puzzled about its significance for days. Like a painter who had striven to find his motif in the landscape and seen it spread out before him in the recursive patterns of nature, Howard had discovered in the passing clouds the connecting theme of his entire intellectual life.

Such were the circumstances of his early school days during the late 1770s and early 1780s, days that passed slowly by as he stared out the windows, wondering what the future might bring. And then, unbidden, it brought the summer of 1783, the year of "horrible phenomena," in the words of Gilbert White of Selborne, when "there was reason for the most enlightened person to be apprehensive."[9] The year 1783 and its sudden bout of climatic change wrought a state of near panic among the populations of northern Europe and exercised a powerful impact upon meteorological thought. It was also to exert a powerful hold on the mind of the 10-year-old Howard. Here was the turning point, the fulcrum of his engagement with the evolving science of weather and climate. He had experienced nothing like it before—an entire summer filled with inexplicable skies—and the excitement that it generated over those weeks and months gave definition to his early ideas.

According to most of the weather journals that have survived to tell the tale, the first few weeks of spring 1783 started out fair and temperate, with an overall balance of overcast and radiant days. Nothing out of the ordinary had yet been reported, and a light fall of snow in March did little to diminish the general expectation of a comfortable summer ahead. As the early months of 1783 wore on, however, a new and unfamiliar pattern began to be widely noticed. Increasingly hot and foggy days were followed by cold, oppressive nights. Something strange, a sort of hovering opalescent fog, seemed to shroud the air more tangibly than the heat that it contained. Instead of dispersing after a few days, as was widely predicted, this grew steadily and oppressively worse.[10]

By June, a sickly haze had seemed to settle everywhere, canceling the sun and deflecting its light from even the clearest of days. Distant hills dissolved from sight as though the visible world was in retreat, while at night the moon and stars could scarcely be glimpsed through the dimming veil of fog. According to Gilbert White, who monitored the changes from his vicarage garden, "the Sun, at noon, looked as black as a Clouded moon, and shed a rust-coloured ferruginous light upon the ground," and he wrote of how the country people began to look "with a superstitious awe at its red, louring aspect." They were right to be fearful: as visibility reduced to dangerous levels, all manner of accidents increased in number over the length and breadth of the land. Horses and cattle grew restless and unnerved, shying at eddies of torpid air or at the thickening swarms of flies that had hatched in food that couldn't be kept fresh in the dispiritingly sultry air; meat rotted in cellars and cool rooms and, according to the nauseated Gilbert White, could only be eaten heavily spiced even on the day that it was killed. The Buckinghamshire poet William Cowper, always alert to any changes in the atmosphere, observed that "dead ducks cannot travel this weather, they say it is too hot for them and they shall stink."[11] He devoted a section of his poem "The Task" to invoking the "dim and sickly eye" of a summer that he described, in a characteristic chain of adjectives, as "portentous, unexampled, unexplained."

Worse, an accompanying sulfurous smell soon settled everywhere, fouling the dry summer air. The smell carried indoors in spite of all attempts to block it out. The atmosphere seemed to revel in its strange contagion. Leaves began to fall dead from the trees in the middle of a June described by Horace Walpole as being as "abominable as any one of its ancestors in all the pedigree of the Junes."[12] With its reeking air, sickly sun, and early, unbidden fall, something had clearly gone very wrong with the season, and the newspapers began to appeal for explanations.

New illnesses began to spread, to the bafflement (and profit) of physicians:

while the very old and the very young complained of breathing and respiratory problems, others learned to live with headaches and nausea, or with escalating bouts of depression. Due to "the unwholesomeness of the air," as the *Bath Chronicle* reported on July 3, "a fever of the putrid kind rages in many parts, which the people term the black fever, and carries off great numbers." Newspapers up and down the land carried updates of the numbers affected, while the "thick disagreeable mistiness" itself remained as unrelenting as it was unfathomed.

But as the haze spread, the people sickened, and violent thunderstorms stirred in the thickening skies, something yet more disturbing was beginning to occur. According to letters and bulletins sent in from abroad, all Europe and half of Asia was undergoing the same dispiriting experience. It seemed as if a blanket of unbelievable weather had draped itself over half the known world.

Then, through the corroded vapors of the evening, came a new and singular vision: the aurora borealis, the northern lights—or at least a sickly and displaced variation. Its beguiling episodes mystified the English and terrified the French, whose southerly latitudes usually denied them all but a trace of such candescence. Night after night, columns of color danced over the veiled horizons, as the night skies of Europe were dazzlingly lit by the streaming solar wind. For many, such as the young Luke Howard and his schoolmates at the Hillside Academy in Burford, the great cinema of light proved a magnificent compensation for all that was strange in the air. They gazed enraptured from the windows of the schoolhouse, while Howard, the sky watcher, wrote descriptions in his journal.

But among the letters of wonder that took their places in the pages of the local newspapers that summer, rumors had begun to appear. Europe had been blighted by major earthquakes and volcanic eruptions and their accompanying legions of dead. Unconfirmed reports of quakes in southern Italy carried estimates of between five thousand and fifty thousand casualties, although as the *London Gazette* pointed out, a shade optimistically, "no distinct accounts have as yet been received." But the tremors did not stop at Calabria; other, more worrying reports appeared of a series of aftershocks that had affected the northern side of the Alpine divide. The earth's upheavals, it was starting to appear, were proceeding in a line heading north.

It would not be long before such vague geophysical worries became a convincing reality. Volcanic explosions in southern Iceland made sudden front-page news that summer, with accounts of the violent formation of an offshore island a matter of alarm and allure. The reports painted a picture of hardened navigators recoiling in terror from the flaming Arctic seas. Then as now, the volcanic conjunction of fire and ice made a lasting impression on the public mind, with its unshakable appetite for earthly powers.

Earth and weather panics were beginning to quicken, and there were more than enough incidents to lend pace to their escalation, including, as the summer wore on, a wave of severe electrical storms that wracked Britain with an unprecedented level of energy. Lightning strikes left a new and sulfurous stench in their wake, as if chemicals had been burned off in the reverberating air. The *Gentleman's Magazine* for July reported a countrywide increase in deaths by lightning, a phenomenon "more fatal, during the course of the present month, than has been known for many years."[13] The elderly witnesses who were sought out for interviews confirmed that conditions such as these had never before prevailed in living or recorded memory. "Britain's Oldest Man" enjoyed a brief, if anonymous, celebrity as he confirmed from the scene of his retirement in Dover that never before in his hundred-odd years had he been unable to make out the opposite shore for so many weeks on end: France had at last been made to disappear from English view.

Here, then, was history in the making, and the age, although horrified by the turn of events, reveled in the sense of occasion. While "1783" may not strike the same revolutionary note as "1789," it was nonetheless a key year of change in the affairs and fortunes of Europe. For what the world had suffered, according to leaders on the front pages of newspapers, was nothing less than a "universal Perturbation in Nature."[14]

It was certainly true that the natural world had been stricken by something enormous but as yet unknown, and the suspicion grew that all these atmospheric and earthly events, whether near or far or merely fast-approaching, were somehow connected and involved. But what connections could there be among these violent convulsions of earth and sky? The new weather episodes were unlike anything known before, and wider forebodings of approaching calamity soon aligned themselves with the recent atmospheric events. Such anxiety could spread with ease, as there was no agreed language for describing the events, no agreed format for taking observations. In spite of past attempts to establish methodologies, no authoritative means of talking about the weather had yet emerged. And though this had been aired as a common complaint, it had never been experienced as a major concern.

But the events of that year of convulsive change had served to expose the meteorological lack, and to expose it more forcefully than was comfortable.

"There are few, I believe, who do not, sometimes, wish that there was more regularity in our climate," observed the *Lady's Magazine* in 1786, for "on the mind of an Englishman, the weather has so powerful an effect, that you find him in different humours in several parts of a variegated day.—In one point of view—(for in many points of view may an Englishman be placed)—he may be called the weather-cock of the creation."[15]

If the history of climate change in Europe had up to then been recorded only on the level of anecdote, this time instruments, and not just weathercocks, were everywhere available to confirm it. And now that the weather had shown itself to be a compelling scientific subject, worthy of renewed and dedicated attention, watching its behavior was akin to witnessing the vast unfolding of a narrative. Eyes were lifted and pens were readied to record the fluctuations of the scene.

Amateurs everywhere were contributing to the growth of scientific culture with a variety of means at their disposal. Zealously descriptive meteorological journals were by far the most common resource, and those kept by Hayman Rooke, a retired but hospitable infantry captain from the Nottinghamshire village of Mansfield Woodhouse, were typical examples of the genre. Publishing them only at the insistence of his friends—or so he claimed—Rooke was at pains to point out that they were not offered to the world as "Philosophical Registers," but merely as approximated tables. These columns of figures, unhelpful even by the standards of the time, were greatly enlivened by enthusiastic anecdotes of lightning strikes and remarkable winds and disturbances of the air, as reported by the stream of passengers who came to visit the captain on the mail coach.[16]

On one particular afternoon in April 1801, as he was looking out his window for the arrival of the mail, he saw a striking display of street clouds (mostly cumulus humilis) massing above his garden:

Remarkable appearance of Clouds on the 13th of April 1801.

The street cloud seen by Hayman Rooke from the window of his study

His description of the "small white clouds in radiated columns" was an impressive delineation of an unusual cumulus effect, and one of the earliest positive identifications of the wind conditions necessary to create it.[17] The clouds, as he reported, remained in their formation for less than a quarter of an hour, but it was long enough for Captain Rooke to complete the sketch for his journal while seated at his well-appointed desk.

Taken on its own, this may seem a marginal moment of amateur science indulged in by an unschooled enthusiast, but it remains nonetheless an outstanding piece of observation—patient, true, and precise. Though far from the atom storms of metropolitan science, rural amateurs such as Captain Rooke were adding to the intellectual texture of an age. For right across the continent of Europe, some of the finest minds were beginning to pump serious thought into the atmosphere: Benjamin Franklin, Jean de Luc, and Pierre-Simon Laplace in France; Horace Bénédict de Saussure in Switzerland; John Playfair and James Hutton in Scotland; Erasmus Darwin and John Dalton in England; and Richard Kirwan in Ireland. An entire generation of European and American researchers, whose primary scientific interests had hitherto been earthbound, renewed their attentions to the actions of the air. And as they worked on the problems thrown up by the strangeness of the recent weather, by these new manifestations of the fickleness of nature, a young man by the name of Luke Howard was following their findings with a scholarly care and attention that belied his modest years.

It was Benjamin Franklin who first proposed that the sudden alteration in the weather of 1783 had a direct geological cause. Might it have been precipitated, as he asked in a paper submitted to the Manchester Literary and Philosophical Society, by an accident of volcanic activity? Four of the world's most active volcanoes (two in Iceland and two in Japan) had recently exploded in deadly sequence.

Whenever a volcano erupts explosively, it sends a plume of gas and debris into the atmosphere. Small eruptions, of which there are usually some fifty or sixty in any one year, produce relatively small amounts of material, which falls to the ground or is quickly dispersed over the immediate location of the blast. The rarer, bigger eruptions, however, can send millions of tons of dust, ash, and sulfur compounds straight through the troposphere and deep into the stratosphere (the second layer of our atmosphere), which are then formed into suspended clouds of aerosols.

Particles of dust are always naturally present in the earth's atmosphere, and they perform a valuable function in screening sunlight either directly, by absorbing radiation, or indirectly, by supplying the condensation nuclei necessary for cloud formation. Through its presence in the circulating atmosphere, the dust veil, as it is known, is believed to contribute to keeping the earth around 3° C cooler than it would otherwise be. Our atmosphere itself is primarily a mixture of gases present in fairly constant relative volumes: 78.09 percent nitrogen, 20.95 percent oxygen, 0.9 percent argon, and trace amounts (less than 0.1 percent) of carbon dioxide, neon, helium, methane, krypton, carbon monoxide, and sulfur dioxide; there are even smaller traces (less than 0.0003 percent) of hydrogen, nitrous oxide, ozone, xenon, nitrogen dioxide, and radon.[18] This breathable gaseous mixture is what we call air, and it accounts for around 98 percent of the total weight of the atmosphere. The other 2 percent is made up of water vapor, the shapes made by which form the myriad clouds, as well as small but significant quantities of solid airborne matter such as dust, sand, pollen grains, sea salt, and smoke from forest fires.

But the volume of the ash and dust clouds ejected by major volcanic eruptions adds an unknown variable to this atmospheric mix, which at its most serious can exercise long-term effects on the global climatic pattern.[19] It is the sulfur content of a volcanic explosion that plays the most significant role, as it forms gaseous clouds of dilute acids such as sulfur dioxide (SO_2) and sulfuric acid (H_2SO_4). Not only can these combine with atmospheric water to fall as acid rain, destroying crops and poisoning livestock, but they can also remain suspended in the atmosphere far longer than volcanic dust or ash itself, serving to block and absorb incoming sunlight, a cause of sudden (and sometimes long-term) cooling. The discolored sun, the atmospheric haze, the early fall of the leaves from the trees, and the spreading sulfurous smell—all these had built a familiar picture by the end of the summer of 1783.

There were a dozen new eruptions that year, two of which are registered as size 4 ("large" or "cataclysmic" on the Volcanic Explosivity Index, or VEI) and three as size 3 ("moderate-large" or "explosive").[20] Compared with 1782 (which had seen three new eruptions, all size 2 or below), 1781 (which had seen four new eruptions, all size 2 or below), and 1780 (which had seen eight new eruptions, again all size 2 or below), 1783 was a major year for volcanic activity, and it was this that drove the changes in the weather.

It had started at the beginning of December 1782 with the moderate eruption of the south flank of Iwaki, a crater on Honshu Island in Japan. A second Japanese volcano, Aoga-Shima, erupted on Izu Island the following April, the first of the year's size 3 blasts. By then a visible tropospheric haze would have formed over most of Japan, although it was unlikely to have been visible farther afield. But in May (and again in August), Asama, a major vent in the Honshu volcanic chain, exploded with a massive size 4 blast, described at the time as "the most frightful eruption on record."[21] The immediate environmental impact was severe, devastating the region around Asama with lava flows, mud slides, and choking deposits of ash, killing over fifteen hundred people outright. Worse, however, was to come. Millions of tons of ash and gases, particularly sulfur dioxide, had been blasted into the stratosphere and remained suspended there in clouds, which proceeded to drizzle toxic rain onto the crops and animals below. All incoming sunlight was blocked for weeks on end. The resulting famine (known as the Tenmei famine) killed more than three hundred thousand people over the next four years. The temperature in the region fell by an estimated 2° C.[22]

In Iceland, meanwhile, on the other side of the world, the offshore island volcano of Reykjaneshryggur erupted in early May, followed by the larger Laki basalt fissure, which began to erupt on June 8 and continued its activity for the next seven months. Terrifying displays of fire fountains and lava flows were staged: in fact, Laki produced the single greatest lava flow of the last millennium (around 15 to 20 cubic kilometers) in addition to the 200 megatons of sulfuric-acid aerosols that were projected into the atmosphere and "widely distributed by the meandering North Atlantic winds."[23] As in Japan, falling ash and acid rain destroyed crops and livestock in Iceland as well as over much of northern Europe. In Caithness in the Highlands of Scotland, 1783 became known as "the year of the ashie." Famine caused by crop failure and the mass poisoning of livestock and fish, in addition to illnesses resulting from fluorine contamination, reduced the population of Iceland by a quarter.[24] The cooling effect was felt immediately in Iceland itself, while elsewhere in Europe the veil of dust and gas served to trap reradiated summer heat, leading to the stultifying conditions described by observers all over the continent.

The following winter, by contrast, was severely cold throughout both the Northern and Southern Hemispheres, due to the long-term global effects of the combined atmospheric haze. Sunlight was screened from the surface of the earth, and Laki's continuous replenishment of the veil of dust over the ensuing months further compounded the climatic impact of the worldwide volcanic sequence.

Every era feels its own climate to be unprecedented; weather has held the front page since the first contrivance of print. It is, as the Reverend John Pointer wrote in his *Rational Account of the Weather* of 1723, "one of the grand Concerns of Mankind. 'Tis what affects all sorts of People, Young as well as Old, Sick as well as Strong."[25] His readers, members of the South Sea Bubble generation, would have had in the back of their minds the other, original disaster, which hit on the night of November 26, 1703. It remains the worst storm in British history, an extratropical cyclone that hammered into Britain from the North Atlantic at more than one hundred kilometers per hour, claiming the lives of around ten thousand people on land and unbelievable sea. They, like their grandchildren in the 1780s, had had nowhere to turn in their hour of need.

As a meteorologist and a man of God, the Reverend Pointer shared their worries, and in the preface to his book he lamented the uselessness of all philosophical schemes and calculations of the weather to date, promising to sift out the "most probable and rational Conjectures" from the rest of the surrounding chaff.

He warned his readers in the meantime to keep away from almanac writers, weather prophets, and other "philomaths" whose "Astrological Cant and Jargon they annually trouble the world with, and with which glittering Starry Notions they are perpetually dazzling and deceiving the Eyes and Ears of the Unthinking Vulgar."[26] Like an early-eighteenth-century Richard Dawkins, the Reverend John Pointer, rector of Slapton (a small village near Towcester in Northamptonshire), was going to sort out the charlatans for good.

Although only an obscure country cleric, Pointer, like so many others of his class and calling, felt himself to be a member of the new generation of amateur inductive scientists, with their instruments and empiricism at the ready. But in spite of the promise of its title and the announcement of its mission to undeceive, Pointer's book ended up as no more than a compendium of anecdotal writings on the weather, gathered largely from scripture and antiquity. The Reverend Pointer was really a weather antiquarian, and his book offered few insights beyond the obvious ubiquity and appeal of his subject. Like Daniel Defoe, whose own awed account of the storm of 1703 surrendered to the conclusion that its ferocity had extinguished the candle of reason, and that only God could be appealed to for an explanation (or a light), he was at a loss to contend with connected ideas of climate and climate change. The system of the weather was bigger by far than any proffered system of thought.

But it was the events of 1783 that helped bring about a change in this approach. The recoil from an individual storm, however severe, diminishes once the dead and the damage have been cleared away. Defoe's book *The Storm*, hastily compiled from a series of eyewitness accounts of the 1703 events, sold only for as long as the topic remained current. Reaction to the weather bouts of 1783, on the other hand, lingered long enough in the mind to disturb, and long enough to ally itself to the wider transformations in thought already under way throughout the Europe of the scientific Enlightenment. The combination of the wider intellectual climate of the times and the strange weather events themselves was to prove a powerful force for change.

So by the time he left school in 1788, Howard had found plenty to think about. The Oxfordshire sky had offered up its mysteries to the eyes of an impressionable boy. His response to what he had seen was only just beginning. Back in London, at his parents' new house in Stamford Hill, then a village to the north of the City, he spent most of his daylight hours working in the garden. There he created, at the end of a serpentine path laid out in gravel and wood shavings, a small meteorological station consisting of a rain gauge, a thermometer, and an inexpensive recording barometer. His mother called the path to the station Luke's Walk, and it was there that his vocation as an amateur meteorologist began in earnest, with twice-daily readings taken and recorded in the pages of the slim pocket journals that he was habitually to keep near at hand. Wind direction, air pressure, maximum and minimum temperature, rainfall, and evaporation—all were enthusiastically noted and logged. Things were beginning to take shape.

It would be a discouragingly short-lived apprenticeship, however. It happened that he was no longer in charge of his own scheme of study, as he had been at the Hillside Academy. After only a few weeks of blissful occupation, in which young Luke communed with the fluctuations of the sun and the rain, his father once more sent him on his way.

The coach journey to Stockport, in Cheshire, took three rattling days along rutted roads, at the end of which Luke Howard delivered himself up as a bound apprentice to Ollive Sims, a business acquaintance of his father's. The elder Howard had calculated that a further period away from London and its distractions would prove the best course of action for his easily deflected son, and his old friend Sims was selected as a trustworthy guardian for the boy. There may have been some justification in his reasoning: clouds, after all, according to Aristophanes, are the patron goddesses of idle men, and as far as the practical Robert Howard was concerned, the sky was a most unlikely source of regular or gainful employment. The boy was to apply himself to more down-to-earth matters and apply himself, besides, in a more down-to-earth location. But for the younger Howard this turn of events came as a cruel blow, especially as his new guardian took to placing ever more unwelcome constraints upon his freedom.

Ollive Sims was a retail chemist and a stern, strictly orthodox Quaker. Unlike the majority of the members of the sect in London, he wore the century-old garb of the early followers, and in his rough black coat and broad-brimmed hat, its band set with the trademark silver buckle, he cut a tall and sobering figure as he purposefully strode around his workshop. There the young apprentice was kept busy from morning till night by the voice of his unflagging master. There was much to be done and little time left over for laughter or lightness, no room for the "youthful levity" that his father warned him to avoid.[27] "Thou must willingly submit to thy Master and Mistress" was the father's response to a complaint of overwork from his son; "it will be to thy advantage to strengthen his hands in business in every way, and not to render the common difficulties of his Life more difficult."[28] The apprentice's duties were onerous, and what with keeping the workshop and laboratories in order, grinding botanical and chemical preparations, rolling out tablets, cleaning out bottles, and stacking the shelves of the shop, he had scarcely a moment to devote to the progress of his thoughts. Sims's shop had recently been built on the site of a demolished pub, a fact that on its own served well as an image of the life-denying sobriety of the Quaker north. Poor Luke's new home life, shared with the Simses in the house next door, was hardly more supportable than work. Meals, like prayers, were taken in silence, and leisure was regarded as a luxury. As the days and the nights merged one into the other, with little to distinguish one week or month from the next, Howard felt his tiredness and frustration mount as his spirits and enthusiasm declined. He detested his new life in the north of England. More than anything else, he wanted to leave to pursue his dream of a life of science in London.

Stockport in the preindustrial 1780s was an uneventful market town with a population of less than ten thousand. It had little in the way of trades and entertainments outside the weekly market, when (according to Luke Howard's obituarist on *The Friend*) it woke up to a hectic sort of animation and then relapsed into a six days' slumber. There was nothing to do, had he ever been given the time to do it, and no one to meet had he been given the time to meet them. This combination of an inactive town and a drab and humorless employer cast a long shadow over Luke Howard's adolescence, as the apprenticeship system had done to the lives of countless young men before him. "Thou hast now two years remaining of thy Term," as his father needlessly reminded him in July 1792, "and it's reasonable to expect in that time an increase of weight will fall upon thee."[29] It was hardly a comforting thought, and his father's ceaseless exhortations by letter to "rise at seven in the morning and to bend thy mind attentively to business, dost thou cheerfully ask leave for every absence and give a faithful account of thy time," did nothing to ease the loneliness, the drudgery, and the unhappiness of his detested provincial exile. It was to go on for six long and difficult years.

The striking feature of Howard's early life was that so much of it had been lived among dominant and inescapable personalities: his father, with his constant improving admonishments by letter; the schoolmaster, Thomas Huntley, with his cane and his textbooks and his strange ideas on the transmission of learning; and Ollive Sims the retail chemist, with his drab-colored clothes, his implacable work ethic, and his sardonic expression, offset only by the silver buckle shining in his broad-brimmed hat. They were all of a type, this cast of characters, looming in authority to form the vigilant panel that had so far sat in judgment on his life. But, like Goethe's Wilhelm Meister, the resourceful young apprentice was not one to be wholly defeated either by his surroundings or by his superiors. In the "short intervals of leisure" that later in life he preferred to try to recall from those early years in exile, he devoted himself to the range of ideas that Huntley's hours of Latin grammar had been unable to suppress.[30] Botany, chemistry, and meteorology continued to occupy his private speculations, and as his thoughts moved forward in spite of their constraints, he discovered in himself the refuge and the rewards of an unusual strength of mind.

Even as he worked in the busy shop in Stockport, even as he submitted to the whims of his employer and the written exhortations of his father, he continued to think about the things that he had seen spread out so vividly over the wide Oxfordshire sky. And as those thoughts took hold, revolved in his mind, and began to assume a vestigial shape, he also began to worry about something else. His father had recently written to him to suggest that he apply for a position at a firm belonging to one of his friends. "If negatived," he concluded to his horror-stricken son, "it will be time to consider what is next best to be done."[31]

Would he ever be allowed to determine for himself the future direction of his life?

Chapter Five

The Askesian Society

Being often *sub dio* in passing to and fro—after sunset as well as by day,—I was brought into a more minute and constant observation of the sky and clouds,—resulting in my work entitled "Essay on the Modification of Clouds, etc.", presented first to the Askesian Society.

Luke Howard, c. 1840[1]

In the summer of 1794 Luke Howard made his longed-for return to London. His six years of exile in Stockport were over, and he had grown more determined than ever to lead a life that had room for his own desires—a life, in other words, that offered space for scientific research. This was not immediately possible, however, and he spent several months working in a junior capacity for a firm of wholesale druggists on Bishopsgate, just to the east of the City. It was the same position that his father had suggested in the letter written two years before. In reality, it was an extension of his apprenticeship, with long hours and little prospect of advancement, although the hardest of his many hard tasks was hiding his impatience from his father. In private he compared himself bitterly to Jacob, who was tricked into serving out two long apprenticeships before making his escape toward the freedom of home.

It was not long, though, before Howard had taken decisive action. He borrowed enough capital—some £2,000—with which to establish himself in trade. The desire to take control of his own affairs had led naturally to the desire to work for himself, ever the goal of the Protestant small-businessman. His plan was to open his own pharmacy, and the establishment he found, at 29 Fleet Street near Temple Bar, was close enough to the heart of the commercial City to answer his needs. The building itself, however, was narrow and uncomfortable, crowded in by its busy surroundings opposite St. Dunstan's church. Howard found himself living in cramped conditions in a small set of rooms above the shop. A laboratory extension, added on to the back of the house, sent chemical fumes up to the rear windows during the day. Smells ached from the very walls of the building. Yet in spite of the discomfort and the personal privations, the location was excellent for developing a business, with plenty of allied trades operating in the vicinity. He soon merged into his entrepreneurial surroundings and began to feel that, at the age of 23, he was at last finding a foothold in the world.

The money he had had to borrow for the venture came, not surprisingly, from his father, and he had to put in long hours of work and application at the shop, not just to earn his living but to convince the still-skeptical Howard senior of the wisdom of his investment. Yet in other respects he was freer than he had been at any previous time in his life. And freedom for Luke Howard, inevitably, meant a renewed engagement with science. In establishing himself on his own, he had created the opportunity once more to indulge his intellectual interests. So on two evenings a week he made his way to the West End to attend chemistry lectures given by the celebrated Irish émigré Bryan Higgins. These were delivered in a set of crowded rooms above Higgins's experimental workshop at 13 Greek Street, Soho, an address that had been a vibrant feature of London life for more than twenty years: the Society for Philosophical Experiments and Conversations, as Higgins grandly named it, branding himself, for good measure, "the didactic experimenter" extraordinary.

Bryan Higgins, a flamboyant, iconoclastic personality, had become something of a cult figure among the younger generation. Like Humphry Davy taking the lecture stand at the Royal Institution, Higgins inspired his audiences with a spirit of unorthodox brilliance, daring them to confront what he saw as the anti-intellectualism of the rising bourgeoisie. He urged his listeners to shrug off the trammels of the past and embrace the new world of the sciences, because science, taught Higgins with a gleam in his eye, was fast becoming the culture of the future.

Science was how the power and influence of the coming century was going to be channeled and exploited. And science, most important of all, was going to be held increasingly in the hands of the young, the disaffected, and the intellectually bold.

At the lectures and discussions, Howard grew close to the group of young, scientifically minded Dissenters who made up the core of Higgins's audience. Artisans and mechanical workers for the most part, or shopkeepers and laboratory workers like Howard, they refused to allow the circumstances in which they found themselves to deflect them from the heroic sense of destiny that was so encouraged by Higgins. They viewed themselves, collectively, as a generation on the rise, and they took warmly to this diffident young Quaker chemist, Luke Howard, who had clearly had to persevere long against the odds to achieve his small share of intellectual autonomy. It was a path that was familiar to most of the new friends he was to meet in Higgin's lecture rooms. It was the bond that would keep them together throughout the events of the following decade.

As a chemist and a Quaker, Howard moved easily into this world of Dissenting science, a world made up of lectures, assemblies, journals, and offprints and imbued with a powerful sense of the future. The brilliance of youthful optimism flared out among the instruments of science and scientific sedition. An infectious enthusiasm spread through the younger elements of these captivated audiences, for they knew that the world was growing ever more reliant upon advances in science and technology. As Higgins had told them, the time had now come for their skills to be recognized as vital and modern. The nineteenth century, after all, was just around the corner. It was the century of promise, they told themselves, the century when science would gain real power for itself and for its young adherents.

Sometimes their enthusiastic sense of the future overruled any foresight they might have possessed. Luke's father once wrote him a worried letter describing an accident involving the grandson of a member of the Barclay banking clan. The youth had been seriously injured by some phosphorus he'd bought from Bryan Higgins's shop following a lecture on the properties of the volatile element:

> the 21/- to the Lectures may not be amiss, only mind what thou meddles with. D. Barclay's grandson some time ago seeing something at Higgins' Lectures which pleased him, was induced to purchase a piece of Phosphorous, this carefully closed in paper he carried in his breeches pocket, till having one day exposed it to the air and replaced it, it suddenly inflamed and burnt his thigh so terribly that he had been in danger of losing Life or Limb, but is now said to be in a fair way.[2]

White phosphorus, a nonmetallic element then chiefly derived from urine or bone, will spontaneously burst into flames when exposed to oxygen in the air. This was a property that Higgins had neglected to mention or, more likely, that young Barclay had forgotten in the days following the lecture. The substance burns so fast and so fiercely that the wound would have been every bit as life-threatening as Robert Howard suggested.[3]

But in spite of his father's warnings, Luke's attendance at Higgins's lecture rooms did him nothing but good, for it was at these gatherings that he first made the acquaintance of William Allen, one of the chief co-architects of the essay on clouds. He was also to be one of the chief co-architects of Luke Howard's future life, for it was this friendship—which lasted half a century, until Allen's death in 1843—that precipitated Howard's gradual transformation from a humble apprentice to a rising scientific star.

William Allen was a natural organizer who was looking for an outlet for his talents. At the time of his first encounter with Howard, he had been working for the past three years at a commercial pharmacy at Plough Court, off Lombard Street, but he was growing impatient with the business methods of its then elderly proprietor, Joseph Gurney Bevan. Allen was of a more ambitious, energetic, and entrepreneurial turn of mind than Howard and had been waiting for Bevan's imminent retirement to seize his chance at a directorship. This, he had decided, was his only possible way forward into autonomy and influence. The pharmacy had been in cautious hands since its founding in 1715, but under the future direction of William Allen and his partners it would become a major international concern, finding world renown as Allen & Hanbury's until taken over in the 1980s by the expanding Glaxo Wellcome.

Allen's plans for the company included the building of a new and larger manufacturing laboratory at Plaistow, then a semirural Essex village some eight kilometers east of London. When he gained control of the company at the end of 1795, Allen offered his new friend Howard the position of manager of the entire Plaistow operation. Howard's of Fleet Street, a one-man operation, had not been doing particularly well and was in no position to compete with the Plough Court pharmacy now that it was to be run by William Allen. Accepting the job would mean moving out of London to live in the Essex village, but it would also mean greatly increased comfort and financial security. This had recently become a more important consideration, for Howard was now engaged to be married.

Mariabella Eliot, known as Bella to her family, was the only daughter of John and Mary Eliot of Bartholomew Close, a much-sought-after district on the south side of the Thames. The Eliots were old friends of Luke Howard's parents, and their prudent daughter, who took some time to respond to the proposal, keeping her suitor in suspense "much longer than was agreeable to his ardent nature," was regarded by his family as an "excellent choice."[4] Once accepted, the delighted Luke was eager for the marriage to be held as quickly as possible, and the wedding was scheduled to take place at the Peel Meeting near Bunhill Fields on December 7, 1796. The day itself, according to Luke's sister, Elizabeth, was "a remarkably cold, dark day, in a frosty season, as uncongenial for the occasion as could well be imagined."[5] The two bridesmaids, cousins of the bride, complained throughout about the bitterness of the season, while the bride herself, as Elizabeth Howard rather cattily noted, "was dressed like a middle-aged Friend" against the cold.[6] The bridegroom, "a slight young man of 24 dressed in light coloured apparel and a triangular hat," was far too nervous to notice and mumbled his way through the wedding responses as best he could. He had never been one for public speaking.

He was also worried about the wedding reception afterward. It was customary for tea to be served to the guests at the new home of the bride and groom, but their only home was going to be above the Fleet Street shop. The living quarters there were too small and cramped for everyone at the wedding to attend; the only solution was to stagger the reception over the course of an entire week of evenings, which is what they did. Different groups of guests arrived at six o'clock for a tea prepared by the bridesmaids, Mary Weston and Ann Sherwood, with everyone huddled in the one room that was remotely presentable for the purpose. It was not so much a drawing room, as one guest noted, as a glorified parlor. How on earth would they manage with children? Quakers or not, they noticed such things, and glances were exchanged among the women.

Mariabella put up with the conditions for a month or so without complaint, but she had come from a wealthy family with a town house in Bartholomew Close and a country house near Pickhurst in Kent. She was rapidly growing disenchanted with living above a shop. Seeing the funny side was one thing; living it every day was quite another. Besides which, she was now expecting their first child (a daughter, Mary, who was born in the following November). It was time to find somewhere else to live. William Allen's offer, therefore, could hardly have been more timely, the financial security of the new position enabling a welcome move into a proper-size family home.

In the meantime, explained Allen, he was also thinking of founding a scientific debating club to meet after hours in the Plough Court laboratory, and he wondered whether Howard would like to join. It would be more wide-ranging than Higgin's chemistry lectures had been, and less focused on one personality. Higgins had recently left London to take up a lucrative position in Jamaica as scientific adviser to the West India Merchants, and his departure created a gap to be filled in the market for Dissenting science. The fortnightly meetings would be held around six in the evening, early enough for members of the audience to go on to meetings at the Royal Society. The venue itself would also be convenient for everyone involved in its organization, located as it was off Lombard Street, the very heart of the Dissenting area of the City.

Cultural allegiances had for centuries drawn themselves directly onto recognized areas of the metropolis, much as they do today, and the Lombard Street Quaker colony was one of the many spiritual knots that gave districts of London their distinctive congregational characters. Like the Huguenot workshops at Spitalfields, the Jewish quarter of Whitechapel, or the Unitarian fellowship at Newington Green, Lombard Street in the City of London had evolved by the end of the seventeenth century into a spiritually demarcated zone. Banks, doctors, chemists, printers, and a plethora of small industrial and manufacturing concerns grew up within a territory marked out quietly by the Quakers for the pursuit of their professional ventures.

They were to flourish particularly well in the banking and pharmaceutical industries, mainly because, unlike most of the others who operated in the shadowy worlds of money and drugs, their word could be taken on trust. At a time of widespread adulteration of medicines, as well as the forging of banknotes and coinage, the Quaker virtues of honesty and plain dealing stood out as life-saving guarantees to their customers. Silvanus Bevan (1691–1761), the founder of the Plough Court pharmaceutical dynasty, had traded not just in goods but in good faith, as did his successors over the coming generations. Quaker clans such as the Bevans, the Darbys, the Lloyds, the Barclays, the Frys, the Cadburys, the Allens, and the Howards were at the center of a distinctive though largely apolitical group that traded within a wider, overtly politicized mercantile environment. Sidestepping any involvement in public affairs, they instead channeled their considerable energies into education, commerce, and scientific research, all of which were encouraged on religious and pragmatic grounds as sources of enlightenment and wealth. This was the outlook defined by Coleridge as "the prime sun-shine spot of Christendom in the eye of the true philosopher," it was the outlook encouraged by Higgins in his lectures, and it was the outlook that would find its greatest expression in William Allen's fortnightly science club.[7]

The Plough Court laboratory in a nineteenth-century photograph

The Askesian Society, as it was named by Allen, was founded in March 1796 amid the blackened apparatus of the Plough Court laboratory. The name was taken from the Greek term *askesis*, meaning "training" or "application," a reminder of the self-improving nature of the project. The three principal founders of the club, William Allen, Richard Phillips, and William Haseldine Pepys, were all young Quakers from the Lombard Street area, and two of them (Allen and Phillips) had worked as apprentices at the Plough Court pharmacy before Allen had taken over the running of the business following the departure of the elderly proprietor. Bevan's other partners had retired early, powerless to resist the force of the takeover. Plough Court was witnessing change.

Number 2 Plough Court was one of many similar buildings that had risen from the ashes of the Great Fire of London. The fire had left the entire Lombard Street area as little more than a knot of smoking ruins, and its replacements were built speculatively as both trading and residential premises. All the buildings in the area were designed so that the proprietor, whether merchant or manufacturer, could occupy the suite of upper rooms.

No. 2 Plough Court, Lombard Street

This was how it had been for the father of the poet Alexander Pope, who was a linen draper (and, more important, a Roman Catholic) during the 1680s. He leased the whole of 2 Plough Court as the premises for his work and for his family. There, Alexander Pope was born on May 21, 1688, in the same building—perhaps even the same room—in which the Askesian Society was to begin conducting its meetings more than a century later. By that time, however, the area was almost entirely populated by Quakers; not long after the birth of the poet, anti-Catholic laws had been passed that prevented their residence within ten miles of the commercial precincts of the City. The Pope family, who had only recently converted to Catholicism, fled to the refuge of Windsor Forest. It was then that the Quaker Bevans took over the evacuated building and began to ply their trade as wholesale and retail pharmacists. This they continued to do for another two centuries to come, with little to disturb their reign over the district. So when the Askesian Society arose, it did so as a tangible expression of the Dissenters' total cultural dominance of the Lombard Street area.

The background and youth of the original members, with their average founding age of 24, were the leading factors in their collective relationship to learning. Since the majority of the Askesians came from Dissenting backgrounds, they found themselves barred under the seventeenth-century Corporation and Test acts from attending grammar schools or either of the English universities, and from holding any form of public or government office. In common with Catholics, Jews, and all other nonadherents to the Thirty-nine Articles of the Church of England, Protestant Dissenters such as Quakers and Unitarians had found themselves painted out of the canvas of public life. Yet they considered that a course of regular study was the only possible means of attaining the scientific or technical knowledge upon which so much future prosperity depended. Since this was denied to them on ideological grounds, they arranged it for themselves instead, inviting as many of their other young acquaintances as they could to join their burgeoning ranks. Schools and science clubs proliferated through the length and breadth of the land. The Askesian Society was just one small island in a swarming archipelago of knowledge.

As at Bryan Higgin's lectures, where many of the group had first met, the natural sciences and natural philosophy were to be the cornerstones of this emancipatory approach to learning, and for some among the more disenfranchised of the urban population, scientific research would itself become imbued with an overtly political meaning. Revolutions in science had already been compared to other revolutions in authority, with the excitement and freedoms of both seen as serving compatible ends. As Erasmus Darwin once confided in a letter to a friend, "Do you not congratulate your grand-children on the dawn of universal liberty? I feel myself becoming all French both in politics and chemistry," a remark that revealed much about the preoccupations of the times.[8] Common attitudes toward human and natural affairs had both been affected by the events of the 1780s. While the French Revolution had come and gone, leaving the political climate of Europe irreversibly changed, the processes governing life and nature were also being unraveled and reorganized along powerful new lines of thought. The year 1783 played its part alongside 1789: the twin years of change in a revolutionary decade.

Erasmus Darwin was not alone in having made the connection between politics and science. The young speech maker John Thelwall, who was to find himself arraigned at the Old Bailey in 1794 on charges of seditious libel, was a fervent supporter of the right to question any received or sacred opinion. Few figures in the 1790s were more radical than Thelwall but, according to the firebrand, when not engaged in political action, denouncing the government in stirring tones, or cutting the froth off a mug of Crown Ale and declaring to the innkeeper that all crowned heads should be served the same way, he was most likely to be found seeking solace in his pellucid attempts to understand the unfolding mysteries of nature:

> I constantly pursue the beautiful phenomena of nature: a pursuit which no accidental changes of the weather can disappoint: for which of these changes does not produce some additional food for science or imagination? If, as now, a sudden cloud envelope the splendid face of heaven, I compare the appearances with the theories which have endeavoured to explain them. If a storm succeeds, I look around for the shelter of some cottage, or little ale house . . . or if matters come to the worst, as in the present instance, I accommodate myself under the shade of some tree, or hovel, and contemplate the operations of nature; see the light mists that had been rarified by the

> warmth of the lower atmosphere, condensed again by the colder region of the air above, and precipitated in lucid drops to the gaping earth, from which they had formerly been attracted.[9]

Thelwall seemed to recognize through his contemplation of the passing clouds a newly wrought model of intellectual freedom that linked the right to roam and gaze up at the sky with the right to ask political questions. Nature and nature's laws were the bedrocks not only of the arts and sciences of the age but of the new political understanding and the social enlightenment that placed freedom of conscience at its heart.

This atmosphere of cultural promise and renewal was breathed in deeply at the early Askesian meetings, where established theories and new hypotheses were ruthlessly tested for strength. The members of the group engaged in discussion and debate, in contrast to the dignified meetings of the century-old Royal Society, where to question a speaker over an assertion was held to be as ill-bred as questioning a tradesman over a bill. Askesian meetings, like coffeehouse debates or John Thelwall's tavern speeches, were characterized instead by a lack of restraint, a lack for which audiences were keen to cram themselves into the Plough Court laboratory. Sometimes so many people arrived that experiments had to be conducted in repeat sittings held over the course of the evening. As at the later Royal Institution, where a number of Askesians went on to become lecturers, the scope of scientific interests was broad, with demonstrations given and papers read aloud on galvanism, ventriloquism, the separation of gases, and mineralogical analysis; and (only once, it seems, and hardly surprisingly) a noisy demonstration was conducted on the manufacture of explosives, by a Mr. Coleman of the Royal Gunpowder Mills.

As part of the metropolitan theater of science, the group of young Askesians shared in the atmosphere of drama and debate, although, amid the noise and the smoke and the sheer mounting excitement, one thing remained abundantly clear: they gathered there not for the sake of talking shop but to expand their intellectual range.[10]

The first paper heard by the assembled society, "On the General Principles of Astronomy," was read by Samuel Woods, the first elected president. It was a characteristically cosmic introduction, and over the course of the coming seasons the speakers and their subjects proliferated: William Allen spoke on "Chemical Attraction," William Phillips lectured on "The Divining Rod," Wilson Lowry talked about "Malleable Zinc," and, of course, the young Luke Howard, who rode in from Plaistow on a borrowed horse, delivered a mesmerizing paper on "The Modifications of Clouds."

Many years later, Howard, himself a product of this age of enlightenment, gratefully remembered the benefits that membership in the club had bestowed:

"Circumstances have long since dissolved this little fraternity, designated by the name of the Askesian society, derived from the Greek ασκησις, *exercitatio*; and I believe many of those who were in the habit of attending will acknowledge themselves indebted to its exercises for some permanent improvement in their scientific character."[11]

Character was not always improved, however. Although Howard's paper was the highest point of the entire Askesian achievement, it was not in fact their highest moment, as a number of the society's earlier meetings had degenerated into drug-taking binges. The experimental ethos of the club encompassed trials of hallucinogens old and new, some of which had surprisingly exhilarating effects. William Allen's diary records an evening in January 1800, at the turn of the exciting new century of science, when William H. Pepys made up a quantity of nitrous oxide (NO_2), "the new gas, from nitrate of ammonia," for distribution at the Askesian Society meeting. At the meeting, which, perhaps not surprisingly, attracted "many visitors," the "gaseous oxide of azote" was passed around and inhaled by everyone present. It had, as William Allen observed, "a remarkably inebriating effect . . . Why your eyes twinkle as if you were drunk."[12]

A few weeks later the Askesians, including Howard, were collectively at it again. According to Allen's diary, "we all breathed the gaseous oxide of azote. It took a surprising effect upon me, abolishing completely, at first, all sensation; then I had the idea of being carried violently upward in a dark cavern, with only a few glimmering lights. The company said my eyes were fixed, face purple, veins in the forehead very large, apoplectic stertor, &c."[13] Oxygen deprivation, known as anoxia, can lead to feelings of intense euphoria, but it is nonetheless alarming to behold. It is the intended outcome of autoerotic strangulation, although it often ends in accidental death. Yet in spite of the scale of its outward symptoms, as well as its increasingly obvious addictiveness, nitrous oxide was declared to be harmless, and its use and abuse went on to become a European craze, particularly after Humphry Davy popularized the gas in a series of pamphlets and demonstrations. Davy himself developed a serious nitrous oxide habit, and, at the height of his dependence, he confessed to inhaling it "three or four times a day," reveling in the "trains of vivid visible images rapidly pass[ing] through my mind."[14] London theaters such as the Adelphi, too, began staging their own "laughing gas" evenings, where members of the audience could line up for a chance to inhale the new wonder drug for themselves.[15]

William Allen, as the director of a leading pharmaceutical company, had a professional interest in sampling any new entry to the expanding *materia medica*; but under Davy's influence he seems to have come close to an NO_2 addiction himself.

He was weaned off the substance through the concerned attentions of his circle of scientific friends. He had had, all in all, a fortunate escape from the horrors of narcotic dependence. "Never get high on your own supply" is modern pharmacology's golden rule, and Allen, as had his friends, had learned it the hard way.

But he was an excellent friend in return, and much of the success of Howard's essay on clouds was due to William Allen's twofold intervention in his life: first the offer of the job out at Plaistow, and then the invitation to join the scientific debating society. Relocation to the outskirts of the city in particular had proved the making of the essay on clouds. Once settled in his new home in Plaistow, Howard rediscovered the time and space for looking up once more at the ceaseless traffic of the clouds. The habit of contemplation, which had all but deserted him during his years with Ollive Sims, renewed itself with gratifying vigor. "In passing between the works and my dwelling," as he recalled, "I resumed the observations I had long been making on the face of the sky."[16] The wide horizons of the Essex marshland granted him the unrestricted use of what amounted to a natural observatory, augmented by the viewing station that he soon had built on the upper floor of his new and spacious home. There, high windows looked out upon the sky in every direction, commanding a view much admired by his circle of scientific friends. And there, amid his books, his young children, and his increasing sense of fulfillment, Luke Howard grew happier and happier. The combination of work, his growing family, and a reinvigorated life of the mind suited him down to the ground. Luke Howard, at last, was in his element.

According to his granddaughter Mariabella Fry, Howard spent most of his spare time up in his workstation, "a sanctum which we might not lightly invade, furnished with books and a perfect carnival of apparatus, scientific and mechanical: these latter reposed in awful glory on the shelves, except now and then, when he would give us an evening lecture on the air-pump, or the electrical machine, and then willing hands collected jars and chains, and receivers, and chairs were set in front of the sideboard, and the household came in to enjoy the marvellous experiments."[17] This is an irresistible scene: Luke Howard the householder, lecturing his captive audience on the secrets of natural philosophy. It was also a characteristically Quaker maneuver to relocate the vibrant theater of science to a scene of self-improving domesticity.

From his upper sanctum of science, Howard was able to watch the unfolding of the weather above the malarial Essex marshes and to monitor the drift of its changes and moods. As at school, clouds formed and fled before his eyes, reviving in him the seeds of an idea that had first occurred to him all those years before. As he pondered the idea, revolving it in his mind this way and that, he also stepped out to view the regular balloon ascents that had become an inescapable feature of both urban and peri-urban life at the turn of the nineteenth century. Balloons were beginning to fill the sky, and the Askesians, unsurprisingly, were keen to experience the uplift for themselves. Drugs were not the period's only technologies of ascent.

Meteorologists had been in possession of a range of instruments for well over a century, and by then, there could have been few parsonages or physicians' rooms in provincial England without a barometer fixed beside the mineral cabinet and the pair of celestial globes. But among their more obvious limitations was the fact that these instruments were resolutely earthbound, even when taken to the highest Alpine summits (as they were by the indefatigable mountain scientist Horace Bénédict de Saussure). The most that meteorological instruments could do—and this includes the umbrella, for which the first patent was taken out in 1786—was to respond to conditions prevailing on earth. But increasingly aspirational attention was being paid to matching conditions in the sky. There was a recognition that weather so far had been touched only from a distance and that a closer form of understanding required a closer form of approach.

An answer to the problem was soon to emerge in the pioneering form of flight. The first balloon ascents, staged in Paris at the end of the atmospheric year of 1783, heralded the advent of a whole new field of meteorological research.

Two French brothers, Joseph-Michael and Jacques-Etienne Montgolfier, initiated the age of the balloon by applying the theory of cloud formation then most widely held to the problem of lighter-than-air flight: the vesicular, or "bubble," theory, which held that "aura," a buoyant form of air rarefied by sunlight, rose in clear bubbles to create the visible floating clouds. If enough of it was successfully trapped, it might be made to keep a passenger afloat. A silk globe filled with the warm vesicular aura would surely rise with more than enough power to lift and transport an aeronaut. Human flight would finally be achieved by imitating not a bird but a cloud. A hot-air balloon, by the terms of this description, was nothing more than a convective current in a bag. The idea (if not the physics) was simple but effective, and after the success of the Montgolfier brothers' first experiments with their aerial balloons, in which chickens, dogs, and goats all survived their pioneering ordeal intact, the time had come for humans to learn to fly among the clouds.

The first manned balloon flight, in November 1783, exerted a profound influence on the European imagination, its impact every bit as powerful as that of the first space travel nearly two centuries later. When the Robert brothers' hydrogen balloon ascended from the Tuileries Gardens in December 1783, a crowd of some four hundred thousand people, half the population of Paris, turned out to watch: it was the largest gathering the world had ever seen. For who would not be moved by the realization of the long-dreamed-of prospect of human flight? Not Jacques Alexandre César Charles, one of the Robert brothers' designers, who is best known today as the framer of Charles's law (which states that the volume of a gas at constant pressure is directly proportional to its absolute temperature). Charles was an ardent balloonist from the start, and his early enthusiasm was more than rewarded when he became the first man to see the sun set twice in one day.

It was a hazy winter's evening in 1783 when Charles accompanied the Robert brothers during one of their earliest ascents. The atmospheric pollution that had marked the entire summer had been dwindling only slowly, and the sunsets remained spectacular throughout the autumn and winter months. Charles was greatly looking forward to the flight. The silk globe of the balloon had been carefully filled with hydrogen gas, generated by throwing sulfuric acid onto a canister of iron filings, and Charles was there to oversee both the safety and the effectiveness of this potentially lethal method. The swaying bag had to be filled to capacity before it could be released from its moorings into the air, and maintaining control of the inflated balloon made liftoff a hazardous undertaking. But once under way, the flight was smooth, lasting two hours, during which time they ascended to a height of more than three kilometers, from where they watched the miracle of a flaming sunset from their floating aerial observatory. The view was phenomenal, and as soon as they touched down, the ecstatic Jacques Charles jumped from the basket, crying out to the assembled onlookers his newly sworn creed: "I care not what may be the condition of the earth—it is the sky that is for me now. What serenity! What a ravishing scene!"[18] He demanded to be allowed a second flight that same evening, which he underwent alone. He rose even higher on the second occasion, bringing the sun back into view, and he stayed aloft until he watched its second setting, ravished by the sight, "hearing himself live," as he was later to express it to his friends.[19] When he finally relanded, safe and exhilarated, he emerged from the basket more rhapsodic than ever, with the image of the twinned sunsets, viewed from the vantage of the soaring balloon, scored indelibly onto his mind.

In spite of ballooning's being a French invention, it did not take long for the aerial enthusiasm to cross the channel, and 1784 turned out to be the British year of the balloon. Manned ascents in Edinburgh, London, Bristol, and Oxford were all witnessed by delirious crowds, and the spectacles were celebrated and satirized in poems, songs, and plays. For once it hardly seemed to matter to British pride that the balloon pioneers were all foreign. Vincent Lunardi's famous hydrogen balloon, which had risen from the Artillery Ground in August 1784, was exhibited at the Pantheon on Oxford Street, where both the "Aironaut" and his machine were hailed as heroes of modernity. The technology of the future had arrived for all to see.

Even the nervous Gilbert White of Selborne, recovered from the ordeals of the previous year, organized watching parties in Hampshire to look out for the passing of Jean-Pierre Blanchard as his aircraft made its stately way across the skies of the English countryside. When it finally came into view, White was characteristically struck with fear for the safety of the passengers, who would surely be "lost," as he worried in a letter to his younger sister Anne, "in the boundless depths of the atmosphere."[20] Here was a version of the fear that the early mariners were said to feel at the prospect of a voyage over the horizon, updated for a modern technology.

The legion of enthusiasts, by contrast, maintained that balloon technology had opened up a new era of heroism and scientific possibility. "*Minerva* shall the tale declare, how *Blanchard* drove thro' Clouds and Air," as one poet promised, while another, Mary Alcock, went further, suggesting with a certain poetic license that anyone with a ticket to fly could now equal or outrun the greatest achievements of the Newtonian age:

> Alas poor *Newton*! Late for learning fam'd,
> No more shall thy researches e're be nam'd;
> For greater *Newtons*, now, each day shall soar,
> High up to Heaven, and new worlds explore.[21]

The promise offered to meteorological research by the new machines had been exploited from the very beginning. The first of the unmanned balloon flights staged by the Montgolfier brothers at Versailles had sent a recording barometer up with the terrified livestock, whose aerial adventures had been watched by a group of eminent astronomers from the safety of the palace observatory.

The following year an American researcher, John Jeffries, paid the small fortune of one hundred guineas for the honor of joining the great Blanchard on one of his London ascents. Jeffries took with him a specially constructed barometer, the scale of which had been amended to show pressures as low as 18 inches. He also took with him a set of glass vials on behalf of Henry Cavendish, a reclusive English aristocrat and natural philosopher. Cavendish had already taken air samples from the summit of Hampstead Heath but wanted samples of even higher air with which to further his researches on the gaseous composition of the atmosphere. As he was not prepared to expose himself to the curiosity of a crowd of strangers by making a public ascent in a balloon, he had his assistant ask Jeffries to take the vials up with him. The vessels had been filled at ground level with distilled water, and the American was instructed to empty and stopper them at measured heights during the 3-kilometer ascent. He carried out his instructions to the letter, and the experiment, which proved remarkably successful under the circumstances, was the first to demonstrate that the lower half of the atmosphere contains slightly declining levels of oxygen at an ambient concentration of just under 21 percent.[22] The accurate findings of the experiment indicate how immediately satisfactory the balloon proved as a meteorological tool. Jeffries, it need hardly be added, just like Charles the year before, became an enthusiastic convert to ballooning and went on to accompany Blanchard (who by then was a world celebrity) on the first aerial voyage across the English Channel, in January 1785. The bundle of letters that he took with him on the flight became the world's first airmail delivery.

Balloons were here to stay, and by the time the first ballooning novel appeared, in 1786, they had established themselves as a permanent part of the scientific and spectacular furniture.[23] Because balloons were the products of a newly awakened interest in the atmosphere, which had been provoked by the weather events of 1783, Luke Howard greatly enjoyed the somewhat un-Quakerish spectacle: one of his pocketbooks records a visit to the Mermaid at Hackney, to see Sadler's balloon ascent on August 12, 1811.[24] James Sadler, a fellow chemist and confectioner turned aerial showman, made dozens of ascents in his newfound career and once had to be rescued from the sea off Liverpool during a failed attempt to float across the St. George's Channel. Balloonists worldwide soon found themselves competing for record feats of endurance, the greatest of which, the first nonstop balloon flight around the world, was finally achieved in March 1999, when Brian Jones and Bertrand Piccard completed the circumnavigation, a trip of 46,500 kilometers, in just over nineteen days. The pair had realized a heroic fantasy that had gripped the imagination of the world for more than two hundred years.

Garnerin and his parachute over Grosvenor Square, London

The whole phenomenon was taken a stage further in London in September 1802, when André Jacques Garnerin, a charismatic French former prisoner of war, parachuted to the ground from a hydrogen balloon at 6,000 feet, landing badly shaken and airsick in a field near Grosvenor Square. Like all balloonists of the age, he used a barometer with which to gauge his height: when it fell to 23 inches on the Paris scale, he jumped, trusting his life to the accuracy of his instrument as well as to the safety of his prototype parachute.

Despite the evident flaws in his parachute's design (the basket swung wildly during the ten-minute drop), he went on to make a series of popular and highly profitable descents over the balloon-crazed city of London. "Garnerin at first such applause did obtain, that the clouds he resolv'd he would visit again," went the refrain of a comic song published soon after, the title page of which told all: *The Parachute; or, All the World Balloon Mad: A much-admired comic song. Written by Mr. Fox. Ludicrously descriptive of the five aerial excursions made in England by Mr. Garnerin.*[25] There was, it seemed, no limit to Londoners' love of

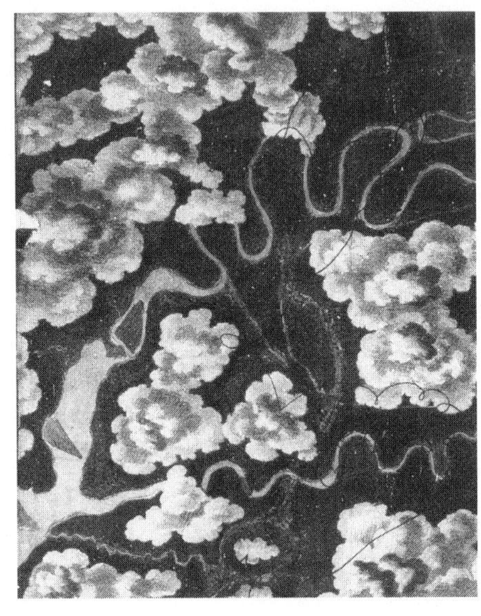

the spectacular, and a man falling to the ground from a great height, swinging suspended and sick from a sheet of white canvas that whistled in the wind as it fell, was just the ticket for an afternoon's outing to the park. Those without tickets watched it all for free from the crowded slopes of Primrose Hill.

The view down from an ascending balloon, or Garnerin's as he floated through the clouds to earth in the autumn of 1802, was the exact corollary of Luke Howard's view as he gazed up from his schoolroom window or from the placid lanes of Plaistow. Here they are, the vaporous clouds, picked out individually above a patchwork of fields. This was a new sight to behold, a downward look at the earth, heralding the advent of a new perspective and a new way of looking at the world.

The world from above the clouds, 1786

Certainly, few in England had seen above the hidden realm of "the Amphitheatre of Clouds" before the invention of aerial ballooning.[26] Mountaineering, in its infancy, was only undertaken en route as a reluctant means of transport. The diarist John Evelyn once encountered mountain stratus on an unforgettable journey through northern Italy in 1644. He had traveled much over Alpine scenery, but this was the view that was to stay with him the longest. For to walk upward through a bank of cloud, to break right through it so as to view it from above, seemed, for Evelyn, to enact a powerful, almost mythic reversal. The impact on his senses was profound:

> As we ascended, we enter'd a very thick, solid, and darke body of Clowds, which look'd like rocks at a little distance, which dured us for neere a mile going up; they were dry misty Vapours, hanging undissolved for a vast thicknesse, & altogether both obscuring the Sunn & Earth, so as we seemed to be rather in the Sea than the Clowdes, till we having pierc'd quite through, came into a most serene heaven, as if we had been above all human Conversation, the Mountaine appearing more like a greate Iland, than joyn'd to any other hills; for we could perceive nothing but a Sea of thick Clowds rowling under our feete like huge Waves . . . this was one of the most pleasant, newe, & altogether surprizing objects that in my life I had ever beheld.[27]

Clouds, for Evelyn, were such an unknown quantity that to pass right through them seemed close to approaching some vast elemental intrusion, breaking into a silent, sacred realm beyond "human Conversation" and thought.

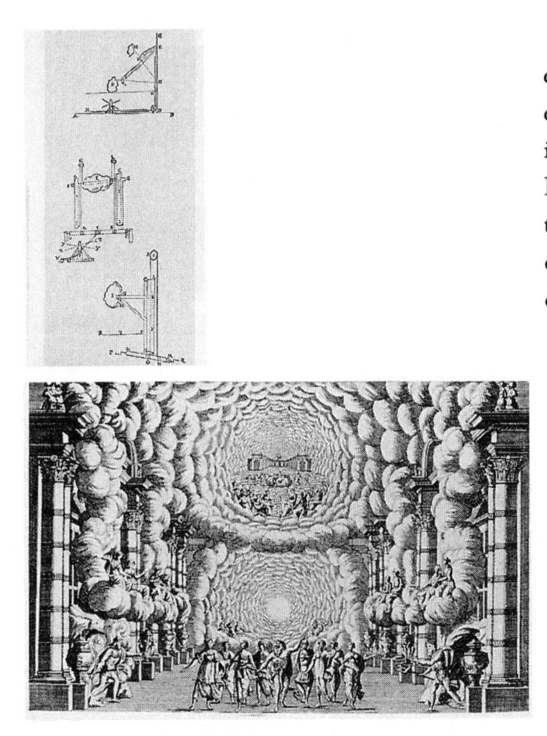
Cloud machines from Venetian opera

Evelyn had encountered something akin to the clouds of the contemporary Italian opera—vast, mechanical, yet strangely mythopoeic structures, caught in a time lapse somewhere between Aristotle and the Enlightenment. For the god-haunted cloudscapes of the early myths had survived in European art, drifting down the continent from legend into opera, via the extravagant entertainments of the Medici. Their wedding celebrations were key events in the power struggles of dynastic Europe, and the demand for Olympian settings and stage designs created the pool of skills that would go on to determine the future direction of Italian musical drama. This multicanopied cloud machine from Giacomo Torelli's designs for a mid-seventeenth-century Venetian opera, for example, shows Apollo's palace complete with an orchestra and corps de ballet, swung high in the air like a visionary city. Nicola Sabbatini's treatise on theatrical machinery shows how it was done. Great wooden arms worked by ropes and pulleys moved the rolls of painted cumulus into position, as rapturous applause from the audience swelled to greet each increasingly fantastic (and power-confirming) statement of a baroque worldview that billowed onto the stage to the sound of trumpets. This was the realm of the gods on earth, and the theatrical image of heaven as an extraterritorial mantling of cloud was, like opera itself, a blend of Aristotelianism and Old Testament cosmology set to the worldly music of the spheres. The mechanics of its delivery, or rather, the invention of clouds, required the attentions of Europe's greatest choreographers and carpenters as well as the fruits of the world's greatest political and musical minds, from Emilio de' Cavalieri, the leader and chief composer of Duke Ferdinand's orchestra, to Bernardo Buontalenti, the greatest set designer of his (and perhaps any other) century. They were a serious matter, these haunts of the gods, and cloud machines took pride of place among the complex elements of the large-scale, gravity-defying set designs that ushered in the birth of the opera.[28]

A century or so later, the secrets of the skies still hovered unobserved, but now there were new technologies helping to reveal them. The times were proving propitious for the study of the atmosphere, and meteorologists embraced the topic and its toys with the delight that characterized the age. New observations, new insights, and new theories were emerging from these growing opportunities for study.

Garnerin and his parachute had been both a sign of the times and a portent of what was to come, for less than three months after his historic descents came the evening of Luke Howard's lecture. And he, too, the young meteorologist, with little more than an insight and a list of Latin words, had risen brilliantly, gloriously to the occasion.

Chapter Six

Other Classifications

HAMLET Do you see yonder cloud that's almost in shape of a camel?
POLONIUS By th' mass, and 'tis: like a camel, indeed.
HAMLET Methinks it is like a weasel.
POLONIUS It is backed like a weasel.
HAMLET Or like a whale.
POLONIUS Very like a whale.

William Shakespeare, c. 1600 [Act 3, scene 2]

But Luke Howard was not alone at the turn of the century in working on a classification of clouds. Unknown to him, a rival attempt at naming the clouds had been initiated in France at about the same time by someone who had already made and lost a reputation for himself. Had this attempt been successful, it could well have raised an unwelcome barrier across the path of Howard's success; and the history of nephology would have a different tale to tell. But it was destined to be just as unsuccessful as all earlier efforts to produce written formulations for the clouds.

For there had already been a number of attempts to list and classify the clouds in the centuries before Luke Howard's evening lecture at the Askesian Society in December 1802, but none of them had left a permanent mark on the scientific or linguistic landscapes. All had faded into the particular obscurity that is reserved for local or fleeting interventions. And though each can be seen to have had its temporary merits, and to have answered some of the immediate needs of its times, it is also not difficult to understand why these efforts had gone on to be unnoticed and forgotten.

Clouds themselves, by their very nature, are self-ruining and fragmentary. They flee in haste over the visible horizons to their quickly forgotten denouements. Every cloud is a small catastrophe, a world of vapor that dies before our eyes. So how, when it has gone, and not a trace of its provisional existence remains, might it be registered as anything other than a cursory sign in the sky? And as long as clouds, for the poetic imagination, stood as ciphers of a desolate beauty, gathering in apparently random patterns only to disperse with the movements of the wind, how could they ever be imagined as a part of nature's continuous scheme? What could there be to a cloud in the sky beyond a vague metaphorical allure?

Like all fugitive effects, clouds require the presence of a validating observer. Yet since no one can see the same cloud twice, and no one cloud can itself be seen twice, there are many more clouds than there can ever be observers. As Jonathan Swift complained in *Tale of a Tub*, a despairing satire on organized religion that was first published in 1704, "to fix upon particulars, is a task too slippery for my slender abilities":

> If I should venture in a windy day to affirm to your Highness that there is a large cloud near the *horizon* in the form of a *bear*, another in the *zenith* with the head of an *ass*, a third to the westward with claws like a *dragon*, and your Highness should in a few minutes think fit to examine the truth, 'tis certain they would all be changed in figure and position: new ones would arise, and all we could agree upon would be that clouds there were, but that I was grossly mistaken in the *zoography* and *topography* of them.[1]

Swift's conclusion is both pessimistic and antiempirical: it is the very extremity of their challenge to the senses that places clouds, like visions, beyond the realm of discussion. If they could not even be known to an individual observer, at least not in any way that was useful or consistent, how might a sense of their forms be communicated? Vague analogy might suffice as far as literary purposes went, but it would never answer the needs of scientific discourse.

And how consciously was Swift echoing an earlier literary model in Antony's complaint to Eros, from a celebrated exchange in Shakespeare's *Antony and Cleopatra*?

Clouds, objects the general, in despair after his defeat at the battle of Actium, are as "indistinct as water is in water":

> Sometimes we see a cloud that's dragonish,
> A vapour sometime like a bear or lion,
> A towered citadel, a pendent rock,
> A forkèd mountain, or blue promontory
> With trees upon't that nod unto the world
> And mock our eyes with air. [Act 4, scene 15]

Antony, like Hamlet, is preoccupied with suicide, and, like Hamlet, he uses the shifting shapes of clouds as a means to demonstrate his own tragic loss of identity. The melancholy reflection rehearsed in the passage, that so much in life that we value is fated to recede into intangibility, would have greatly appealed to Jonathan Swift, the future dean of St. Patrick's and author of *Gulliver's Travels*. Since there was no viable distinction to be made between a reported sighting and an imaginary glimpse, he reasoned, clouds could live as figments in the mind just as fully as they lived as entities in the sky. If "'tis certain they would all be changed in figure and position," what difference did it make what or where they were? Just as for John Evelyn in the Italian Alps, or Hamlet in his father's lonely castle, or Antony outside his tent in Alexandria, they offered little more than a soft impediment to discourse.

So how might they come to be fixed by the attentions of a natural philosopher?

As has been seen in earlier chapters, the idea that our eyes could only be mocked by clouds and air had been disputed on and off since antiquity. Following the example of Descartes in particular, the seventeenth-century rise of Baconian scientific method, with its emphasis on the gathering of material and statistical evidence, was beginning to revolutionize the European approach to both earthly and aerial phenomena. Although Swift and his dismissive Scriblerian contemporaries might have mocked the efforts of the promoters of the new science, the natural world was beginning to be viewed increasingly through their eyes. The universe was coming to be seen as a vast repository of objects and effects to be itemized, cataloged, and understood. New instruments and a new empiricism were equally ready to be pressed into the service of the scientific project. Fixing upon particulars, in Jonathan Swift's impatient phrase, was to be celebrated as the route to the discovery of the general principles of nature. And such principles, it was felt, would inexorably reveal themselves from out of the body of collected evidence.

But most attempts to rationalize the study of the weather, whether on the individual or the collective level, have been dogged by errors and misfortune. When the development of instruments in seventeenth-century Europe led to the first organized attempts to collect centralized weather statistics, the ambition of such schemes required firm guidance and huge resources with which to overcome the many hurdles in their path. But even when rich and powerful patrons arose to fill the financial and administrative need, difficulties remained.

There were few patrons more rich and powerful than Grand Duke Ferdinand II of Tuscany. The grand duke (who, incidentally, liked to dress up as a Turk) was behind the world's first major weather-monitoring project, which ran for thirteen years, from 1654 to 1667, and succeeded in establishing a series of recording stations across the length of northern Italy and over the Alps into Central Europe. Around a dozen stations were established under the scheme, often at preexisting astronomical observatories outside the major cities. The stations were equipped from Ferdinand's opulent headquarters in Florence with sets of similarly calibrated instruments and a standardized procedure for recording the observations to be made: local barometric pressure, wind direction, temperature, humidity, and visibility were all logged by observers onto specially printed forms and then dispatched at regular intervals to Ferdinand's Accademia del Cimento in Florence for "analysis," which in practice, of course, turned out to mean filing. Although the filing was exemplary and records, at least for a short period, had indisputably begun, the project was not in other respects much of a success. The dispatches took too long to arrive at headquarters to be of any but historical interest, which was a perennial problem in the centuries before the invention of the telegraph. When the Accademia was eventually disbanded in 1667, through the efforts of a disapproving clergy, the European weather stations, one by one, died a corresponding death. And when Ferdinand himself died three years later, a major force in atmospheric administration had been stilled.[2]

But he had not been the only one to find himself preoccupied by the weather in the 1660s. Robert Hooke, the first curator of experiments at the newly instituted Royal Society of London, was briefly transfixed by the problem of recording the variable transitions of the atmosphere. In 1665, at age 30, he proposed "A Method for Making a History of the Weather" and produced guidelines for the collection of the statistics. According to Hooke's recommendations, wind force, temperature, barometric pressure, and humidity were to be collected and expressed in numbers, while "Faces or visible appearances of the Sky" and its "Notablest Effects" were to be noted and described in words.[3] This information, claimed Hooke, would soon prove as indispensable to natural philosophers as a set of tools to an artisan.

But, like other early attempts to collect weather statistics, Hooke's scheme was fated to come to nothing in the end. Despite his gregariousness and his wealth of connections, Hooke failed from the outset to capture the level of enthusiasm needed to keep the project running effectively. Volunteers, mostly friends of the

A

SCHEME

At one View representing to the Eye the Obſervations of the Weather for a Month.

Dayes of the Month and place of the Sun. Remarkable houſe.	Age and ſign of the Moon at Noon.	The Quarters of the Wind and its ſtrength.	The Degrees of Heat and Cold.	The Degrees of Drineſs and Moyſture.	The Degrees of Preſſure.	The Faces or viſible appearances of the Sky.	The Notableſt Effects.	General Deductions to be made after the ſide is fitted with Obſervations: As,
14 II 12.46 — 4, 8, 12, 4, 8, 12	27 ♉ 9.46. Perigeu.	W 2. 3. 3. W.S.W. 1	9, 12, 16, 10, 7	5, 8, 9	29 1/10, 29 1/8, 29 1/8	Clear blew but yellowiſh in the N. E. Clowded toward the S. Checker'd blew.	A great dew. Thunder, far to the South. A very great Tide.	From the laſt Q. of the *Moon* to the Change the Weather was very temperate, but cold for the ſeaſon; the Wind pretty conſtant between N. & W. A little before the laſt great Wind and till the Wind roſe at its higheſt, the Quick-ſilver continu'd deſcending til it came very low; after wch it began to re aſcend, &c.
15 II 13.40 — 8, 4, 6, 10	28 ♉ 24.51.	N. W. 3. 4. N. 2. 1	9, 8, 7	28, 9, 10	29 1/16, 29	A clear Sky all day, but a little Checker'd at 4. P. M. at Sunſet red and hazy.	Not by much ſo big a Tide as yeſterday. Thunder in the North.	
16 II 14 37 — 10	N. Moon. at 7. 25' A. M. II 10. 8.	S. 1	10	1 10	28 1/2	Overcaſt and very lowring.	No dew upon the ground, but very much upon Marble-ſtones, &c.	
&c.	&c.	&c.	&c.	&c.	&c.	&c.	&c.	

Robert Hooke's weather chart of 1665

curator dragooned into participating in the scheme, soon grew weary of the task of submitting endless observations logged onto endless forms that they knew would be stored, unread, until a use might one day be found for them. This may well have been an example of lofty and purposeful science, Baconian to a fault, but to many it just seemed like pointless administration. The instruments and the forms may have been impressive, but what, and who, were they actually for? Odd individuals clearly enjoyed the ritual element of the daily task of recording, but most quietly gave it up to follow more obviously amenable outdoor pursuits, such as potholing, botany, or cloud watching.

Even Robert Hooke himself, the brains behind the project, gave up his daily meteorological journal after only a few months of the chore. He had made the same glad discovery as his friends: life was too full to spend valuable time writing weather reports on behalf of the Royal Society. He had also come to the same realization that Luke Howard would nearly a hundred and fifty years later: that a meteorologist who attended only to his instruments may be said "only to examine the pulse of the atmosphere."[4] Science was unsuited to the study of nature if it failed to develop an observational language.

But cloud watchers might have had more to gain from Robert Hooke had they stuck with him more loyally, for he had also proposed a solution to this lack. He devised a language with which to describe "the faces of the sky"—faces that were "so many, that many of them want proper Names."[5] In proposing that the problem of weather lay more in naming it than in the task of mutely recording it, Hooke knew that he was on to something interesting. What was needed was a language of weather that would live beyond the limitations of the printed form, but a language that could nevertheless be expressed within those same limitations.

Hooke had plenty of suggestions for filling the linguistic gap, starting, as one might expect, with the special case of clouds:

> Here should be observed, whether the Sky be clear or clouded; and if clouded, after what manner; whether with high Exhalations or great white Clouds, or dark thick ones. Whether those Clouds afford Fogs or Mists, or Sleet, or Rain, or Snow, &c. Whether the underside of those Clouds be flat or waved and irregular, as I have often seen before thunder. Which way they drive, whether all one way, or some one way, some another; and whether any of these be the same with the Wind that blows below.[6]

Hooke's proposal then led on from the cloud descriptions to a wider discussion of the sky, with an entire vocabulary of recommended terms for catching the variety of its vaporous states:

> Let *Cleer* signifie a very cleer Sky without any Clouds or Exhalations: *Checker'd* a cleer Sky, with many great white round Clouds, such as are very usual in Summer. *Hazy*, a Sky that looks whitish, by reason of the thickness of the higher parts of the Air, by some Exhalation not formed into Clouds. *Thick*, a Sky more whitened by a greater company of Vapours. Let *Hairy* signifie a Sky that hath many small, thin and high Exhalations, which resemble locks of hair, or flakes of Hemp or Flax: whose Varieties may be exprest by *straight* or *curv'd*, &c. according to the resemblance they bear. Let *Water'd* signifie a Sky that has many high thin and small Clouds, looking almost like a water'd Tabby, called in some places a Mackeril Sky. Let a Sky be called *Waved*, when those Clouds appear much bigger and lower, but much after the same manner. *Cloudy*, when the Sky has many thick dark Clouds. *Lowring*, when the Sky is not much overcast, but hath also underneath many thick dark Clouds, which threaten rain. The signification of *gloomy, foggy, misty, sleeting, driving, rainy, snowy*, reaches or racks *variable*, &c. are well known, they being very commonly used.[7]

Here was the first true attempt in Western science to mold a descriptive vocabulary to fit the fleeting appearances of the sky. "Water'd," "Waved," "Cloudy," and

"Lowring" were intended to be agreed upon as fragments of an empirical language. These terms themselves may in practice have been too loose and unsystematic to offer the precision that Hooke was after, but the attempt itself marked a definite shift in attitude. The classifying outlook ushered in by Baconian and Cartesian forms of science proposed that everything in nature, even the clouds and the faces of the sky, could be fixed by the hand of descriptive science. Hooke was "secularising the sky," as the filmmaker and anthologist Humphrey Jennings once put it, "making out of it the subject matter for the new science of meteorology."[8]

But Hooke's disengagement with meteorology was characteristically rapid. According to the *Dictionary of National Biography*, he "hurried from one inquiry to another with brilliant but inconclusive results," and within months of publishing his strictures on the weather his researches had moved from microscopes to diving bells, from diving bells to pendulums, and from pendulums to the vibrations of musical notes. He had almost entirely forgotten about the weather. It is clear that the word *specialist* was not to be found in Robert Hooke's vocabulary.

The idea of an agreed meteorological language with which to comprehend the actions of the air was revived a number of times over the following century, but none of these efforts came anywhere close to a successful outcome. It was only when the

Societas Meteorologica Palatina first convened at Mannheim, under the guidance of the Prince-Elector Karl Theodor, that any real progress was made.[9] Theodor, who offered up his castle at Mannheim for use as the society's headquarters, was, like Grand Duke Ferdinand before him, a passionate observer of the weather, and he was equally determined to understand it through the collection of widespread data. His society's founding purpose was to predict future changes in the weather by studying its long-range patterns and movements and thereby uncover its secret processes. The society's sophistication and resources promised at first to match its ambitions. From a core of a dozen stations in Central Europe, the network spread to fifty or more locations, "extending from Siberia across Europe to Greenland and eastern North America."[10] Dispatched to observers as far afield as Buda, Würzburg, Rome, Spitzbergen, Stockholm, La Rochelle, and Lower Silesia was a set of recording instruments: barometer, thermometer, hygrometer, rain gauge, wind vane, and electrometer, all with standardized scales and firm instructions for their use, written in heightened scientific Latin by the prince-elector's court chaplain, Johann Hemmer. All in all, it was a major feat of diplomacy and logistics.

It was marked, as were all similar projects since the seventeenth century, by a relentless quest for a language of aerial description. The personnel working at the European stations needed to communicate their findings and observations to one another, and schemes of symbols and abbreviations emerged as a result. The one

ABBREVIATION OF SYMBOL	SPECIFICATION
a	*White clouds*
cin	*Grey Clouds*
n	*Dark Clouds*
l	*Orange-yellow clouds*
r	*Red clouds*
t	*Thin Clouds*
sp	*Thick Clouds*
fasc	*Streak-like clouds*
rup	*Rock-like clouds*
lact	*Disc-shaped clouds, of milky appearance*
⋛	*Layered Clouds*
ɯɯ	*Gathering clouds*

The Royal Society of Mannheim's classification of clouds

that was developed for cloud sightings is of particular interest here. It updated Robert Hooke's century-old definitions, while predating Howard's by a mere twenty years.

The terms and symbols, descriptive on their own, could also be used in conjunction. *Cin.sp*, for example, referred to dense gray clouds, while *fasc.l* denoted clouds of a streaky orange. Like Robert Hooke's earlier terminology, this system lent support to the slowly evolving idea that the clouds could be meaningfully named and described. But it was the combination terms that hinted, significantly, at the novel idea of modification. Clouds could move, one into the other, and a terminology was needed to reflect that. The society's solution, though partial, was a bold one.

In the course of his research into the history of meteorology, Luke Howard acquired a copy of the *Ephemerides*, the register of observations published by the society, and so would have been aware of the earlier classification, although only after he had published his own. The thought must have struck him that, had the efforts of the society been allowed to continue, the history of meteorology in Europe may well have turned out differently. But the ill-fated society was disbanded under violent circumstances in 1795, when the city of Mannheim was invaded and destroyed by the French Revolutionary army.

As mainland Europe convulsed under waves of war, invasion, and terror, it would fall increasingly to Britain and Ireland to lead the way in the promotion of science. Napoleon and his armies were responsible for halting much of scientific and cultural value besides the activities of the Mannheim observatory. Many such episodes characterized the times, but a particularly sad one concerned the well-known figure Jean-Baptiste Lamarck, one of whose many areas of scientific interest was closely bound up with Howard's. It was far closer, in fact, than either of them might have imagined, for Lamarck had spent the first few months of 1802 preparing a classification of clouds. Luke Howard, it seemed, was not the only one working on the problem.

European science at the end of the eighteenth century was much taken up with its encyclopedic projects of classification and naming. Travelers were returning from voyages of exploration laden with cargoes of new and unknown specimens, and the search for meaningful patterns in nature's array was coming increasingly to preoccupy the scientific mind. Earlier in the century the Swedish botanist Carl von Linné (known to the Latinizing world as Linnaeus) had introduced the system of binomial nomenclature to natural history, whereby every identifiable kind of organism could be designated by a pair of Latin names, the first denoting the genus to which it belonged, the second denoting the species: *Ardea cinerea*, for example—the gray heron; or *Canis familiaris*—the domestic dog.

Although hugely influential from the start, the logical structure of the Linnaean system implied, problematically for some, that species were permanent, distinct, and immutable, an idea that was already attracting a certain amount of resistance. Other natural scientists, such as Georges-Louis Leclerc, comte de Buffon, soon sought ways to overcome the rigidity of Linnaean classification, which they saw as the flawed result of an overly abstracted approach. For Buffon, nature did not break down into fixed species, as Linnaeus had seemed to suggest, but instead formed a sequence of variously connected individual beings. The task of natural history would be to seek relationships and modifications between these chains of individuals rather than describing the fixed essences of species.[11] Lamarck, as an employee of the well-funded Muséum d'Histoire Naturelle, was more than familiar with both kinds of outlook, and it might seem, on the face of it, perfectly appropriate that he should choose the latter approach through which to attempt his rival classification of clouds. But strange as it might seem, Buffon's system, distinguished by its emphasis on change and mutability, was never to achieve the authority of naming enjoyed by Linnaean organization, even in an area of obvious change like nephology. Lamarck, as so often in his life's allegiances, had made an unlucky error of choice.

Jean-Baptiste-Pierre-Antoine de Monet de Lamarck (1744–1829) can be ranked among the greatest of the unluckier figures who populate the pages of the history of science. Throughout his career, his work repeatedly came off the worse against more successful, more flexible contributions than his own. The story of the failure of his cloud classification forms a typical episode in his unfortunate life.

Having fought (and been injured) as a young man during the Seven Years War, Lamarck arrived in Paris in the 1760s and was soon absorbed in the scientific atmosphere of the time. Like those of Howard in England, Lamarck's main early interests lay in the fields of botany and meteorology, and in 1799 (by which time he had found employment as a curator at the Muséum) he published the first of what was to become a series of annual meteorological digests, the ill-fated *Annuaires Météorologiques*. The volumes consisted largely of elaborate astrological descriptions of the effect of the moon and the planets on climate, linked to weather forecasts for the entire year ahead. These forecasts were not to be regarded as "opinion," according to their wildly overconfident author, but as "fact."[12] Almanacs and forecast books had been a popular literary form for centuries, and Lamarck's annuals sold well, despite the predictable failure of his forecasts. Proved consistently wrong, the author claimed merely that he was "unlucky," and passed the blame on to the lack of statistical material available for him to consult.[13] He had published an appeal to his contemporaries for weather observations to be logged and submitted on a standardized form, but as was usual in the nature of these things (as Robert Hooke had discovered more than a century before), nobody cared to reply. Lamarck, true to character, suspected that a plot had been orchestrated against him by jealous and contemptuous colleagues. Lamarck was never an easy person to get along with. Constitutionally unable to win friends and influence people (a necessary skill in both pre- and postrevolutionary France), he was continually stalled in his career by his capacity to alienate and antagonize his supporters. Even his long-suffering friends were dismissive from the start about the worth of the meteorological almanacs, Louis Cotte going as far as accusing them in print of "hampering the progress of science."[14]

Yet the *Annuaires* were intended to foster genuine scientific research as well as to carry prognostics of the weather, and in the third volume, published in 1802, Lamarck announced an important new project aimed at "the clarification of meteorological phenomena."[15] The project was intended to produce, among much else, a workable classification of clouds based on Lamarck's penetrating observation that "clouds have certain general forms which are not at all dependent on chance but on a state of affairs which it would be useful to recognise and determine."[16]

This insight itself was very close to Howard's, and had Lamarck pursued it in a different way, he would now have a chapter to himself in the history of clouds and climate. But though he had made the crucial recognition that every cloud could be described through a limited number of basic forms, he preferred, following Buffon, to view clouds in terms of individual entities rather than as members of Linnaean species. Any language of clouds to be devised would therefore need to reflect this outlook, and there were hundreds of clouds to describe. Lamarck approached the problem by proposing initially broad families of clouds distinguished only very loosely by appearance. He had no idea how many cloud families he might eventually have to devise, and he proposed the following five as the first in a series of installments:

En forme de voile (hazy clouds)
Attroupés (massed clouds)
Pommelés (dappled clouds)
En balayeurs (broomlike clouds)
Groupés (grouped clouds)

Aspects of these loosely descriptive terms are echoed in Howard's own terminology, such as the "broomlike" cirrus or the "dappled" altocumulus, but in common with earlier attempts to name the clouds, Lamarck's terms were marked by a lack of specificity and precision. They could pinpoint neither a cloud itself nor its capacity to change.

Lamarck continued to publish additional notations in subsequent editions of the yearbook, introducing other terms, such as *nuages moutonnées* (flocked clouds), *nuages en lambeaux* (torn clouds), *nuages en barres* (banded clouds), and *nuages en coureurs* (running clouds), that again lacked the qualities necessary for the establishment of a new scientific language of clouds. He attempted later still to introduce descriptive secondary adjectives, such as "isolated," "obscure," and "undulatory," believing that it was only a matter of time before every cloud species could be identified through the compilation of a full descriptive index of their forms.[17] Lamarck was right in believing that this would happen, but he was right for all the wrong reasons.

Like earlier attempts to classify the clouds, such as Robert Hooke's in the 1660s or the Societas Meteorologica Palatina's in the 1780s, Lamarck's efforts had merely produced another battery of descriptive terms that made broad reference to secondary characteristics of shape, color, and texture. All shared the common failure to understand cloud formation and transformation, manifested clearly in the terms they devised. All these terms, whether "hairy" or "wav'd" (from Hooke), "streaky" or "milky" (from the Societas), or "hazy" or "torn" (from Lamarck), had been developed to assist in the compilation of general descriptive accounts of the state of the sky, and even though Lamarck had attempted, in this instance, to formulate a dedicated typology of clouds, his terms all shared in the same flawed looseness of expression.

Lamarck's terms were couched, in fact, in a strangely pastoral form of French, reminiscent of the language of the revolutionary calendar, the product of an avowed attempt to decimalize the passing of time. The new calendar had been instituted in July 1793 by the National Convention, which also declared September 1792 to be, retrospectively, the beginning of Year One, with all future years to be numbered and named accordingly. This break with Gregorian time was intended to signal a break with the entire ideological history of pre-revolutionary France, a break emphatically reinforced by the abolition of the monarchy in the same year, as well as the declared rebirth of the former nation as a republic. The birthday of this new republic was fixed as September 22, the autumnal equinox, a suitable moment for the new France to be deemed wholly autonomous in both space and time. It was to be a new beginning, with the country and its identity, through the reordering of its calendar, given back symbolically, if confusingly, to the people.[18]

The mathematical and bureaucratic skills needed to work out the new divisions of the year were found in the person of Gilbert Romme, assisted by a number of astronomers, while the task of thinking up the names for each of the new months and days fell to the popular playwright Fabre d'Eglantine, a friend and follower of Danton. (An unlucky friendship, as it turned out—they were guillotined together in 1794.)

In the new system, the year was to be divided along lines derived from the ancient Egyptian calendar, into twelve equal months of thirty days in length, with the remaining five days (or six in a leap year) dedicated to festivals and added to the end of the final month. Each month would also be given a new republican name based on its primary meteorological or agricultural feature. The chosen symbolism was apt: the vast majority of the French population remained employed in rural agriculture, and the ideal of nature had as powerful a grip on the Revolutionary Convention as the ideals of freedom and the fair distribution of food had had upon those who had stormed the Bastille in the summer of 1789.

The months of the revolutionary year (with New Year's Day on September 22) were divided into seasons marked by the onomatopoeic seasonal suffixes *-aire, -ôse, -al*, and *-or*, designed by d'Eglantine to help fix the passage of the agricultural year into the language and the hearts of his people.

The French year now ran as:

Vendémiaire, the month of vintage (September 22–October 21)
Brumaire, the month of fog (October 22–November 20)
Frimaire, the month of frost (November 21–December 20)
Nivôse, the month of snow (December 21–January 19)
Pluviôse, the month of rain (January 20–February 18)
Ventôse, the month of wind (February 19–March 20)
Germinal, the month of budding (March 21–April 19)
Floréal, the month of flowering (April 20–May 19)
Prairial, the month of grass (May 20–June 18)
Messidor, the month of harvest (June 19–July 18)
Thermidor, the month of heat (July 19–August 17)
Fructidor, the month of fruit (August 18–September 21)[19]

Here was the calendar expressed as both a dechristianized and a declassicized agrarian schedule, but one that reminded the people of how they lived at the bounty (and the mercy) of the elements.

The terms, imposed on the French by legal decree, were given a mixed reception both there and abroad, with the English producing characteristically deflationary translations: "Wheezy, Sneezy, Freezy, Slippy, Drippy, Nippy, Showery, Flowery, Bowery, Wheaty, Heaty, and Sweety," ran one, although Thomas Carlyle offered a more temperate (but no less preposterous) rendition in his *History of the French Revolution*: "Vintagearious, Fogarious, Frostarious, Snowous, Rainous, Windous, Buddal, Floweral, Meadowal, Reapidor, Heatidor, Fruitidor" ("*dor* being Greek for *gift*").[20]

The reorganized calendar remained in use in France until the end of December 1805, and so provided the background to Lamarck's meteorological ideas, setting the linguistic tone of his naming of the clouds, as well as imposing on it something of the flavor of reform. Much else in France was standardized and decimalized at this time, including weights, measures, currency, and timekeeping, and Lamarck's attempt to decimalize the clouds took its place within the wider expression of the revolutionary love of order. But with its burden of homely, republican adjectives, Lamarck's scheme ended up as more of a parochial miscellany than a real scientific classification. And, just like the unpopular calendar, it was about to be unceremoniously dropped.

Despite its flaws, however, Lamarck's classification was the first to suggest that clouds might be graded by altitude, in tiers designated as high, middle, and low. This was an influential idea, and the International Meteorological Congress that met in Paris in 1896 adopted a version of Lamarck's three height categories that remains in use today: High clouds (such as cirrus), which form at levels of 5 to 13 kilometers; Mid-level clouds (such as altostratus), which form between 2 and 5 kilometers; and Low clouds (such as cumulus and stratus), which typically form at heights of less than 2 kilometers. But in all other respects Lamarck's scheme failed to make any kind of impression, even in France, due principally to the reasons already mentioned. In failing to adopt the Linnaean style of nomenclature, Lamarck placed his clouds outside the international language of taxonomy. Choosing French over Latin only compounded the problem, since clouds, after all, are a global phenomenon—*nuages sans frontières*—and besides, France's immediate neighbors, Britain and Germany, were hardly in a mood to adopt a French meteorological nomenclature at the height of the Napoleonic wars. It would have been tantamount to a linguistic invasion.

But there was another reason for Lamarck's sharp meteorological decline, as well as for Howard's unimpeded rise, which was connected directly with the actions of the warlike Napoleon himself.

Britain and France had been at war since 1793, and although the Treaty of Amiens (signed on March 25, 1802) had been intended to put an end to hostilities once and for all, within weeks it was clear that they would inevitably have to be resumed against the unchecked expansionism of the French under Napoleon. Rumors of Napoleon's colonial plans included the invasion of Britain by means of troopships, tunnels, and the newer technologies of the air. In the end the invasion never got off the ground, but as volunteer defense forces massed along the coast, the country prepared itself for the worst.

So war and revolution dominated the background of both these turn-of-

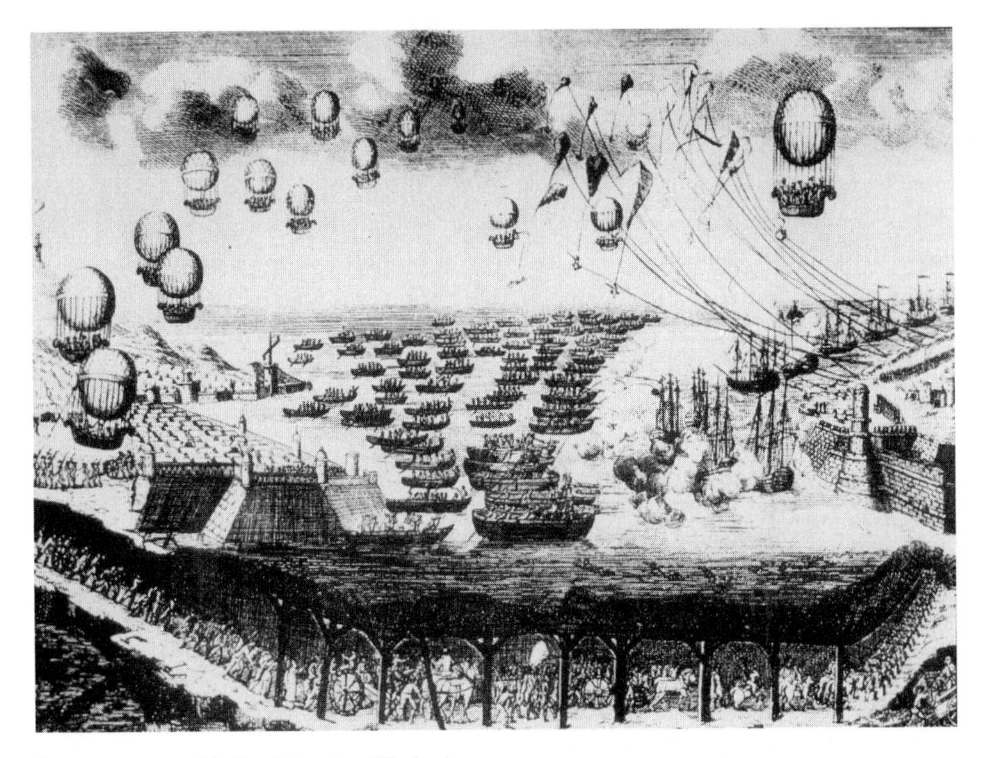

An imaginary view of the French invasion of England

the-century cloud classifications, adding a historical poignancy to their unknown rivalry. The mood of the times was well caught by Captain Hayman Rooke (retired), the amateur meteorologist from Mansfield Woodhouse. In his "Meteorological Register" for 1802–03 Captain Rooke forecast:

> *Remarkable Occurrences.*
>
> The most remarkable one in this year, is a threatened appearance of an infernal and transitory Meteor, which with its diabolical Satellites have already laid waste a great part of Europe; and now in its northern course of devastation, threatens the destruction of this happy Island by storm and fire; but we may hope, under the Protection of Divine Providence, that the Thunder of British Cannons by Sea and Land, will cause this destructive and inflammable Meteor, to burst and end in *Vapour*.[21]

Captain Rooke had reason to be pleased with his military-meteorological metaphors, imbued as they are with the flavor of a period smitten as much with the chauvinism of war as with the onward march of science. The availability of meteorological language for figurative ends is striking, particularly at a time when the consequences of invasion were so feared. Scientific terms and ideas had been feeding into the common language since the Copernican and chemical revolutions (indeed, the very concept of revolution itself was one that derived from cosmology), but this was a more deliberate and knowing piece of wordplay, linking the metaphor of weather to the violent historical agency of war. A century later, in the early 1920s, the figurative precedent was reversed when meteorologists of the Bergen School named the narrow bands between air masses "fronts," after the battle zones of the First World War. Since fronts in meteorology are the buffer zones between moving masses of warm and cold air, and are responsible for the violence of much of our weather as well as the restless energy of the clouds, the imagery of battle was well chosen.

And Captain Rooke chose well, too, in viewing Napoleon as a system rather than as a man. Bonaparte (who declared himself emperor in 1804) was a complex character: a voracious reader of romantic fiction who nevertheless took a dim view of astrological writing. Almanacs, ephemerides, and other forms of celestial and proverbial literature had long been widely read in both Europe and North America. *Old Moore's Almanac*, first published in 1700, was, by the end of the century, selling an impressive four hundred thousand copies a year. The almanac writers' regular attempts at weather prediction were a hangover from the ancient astrological domination of the study of both celestial and sublunary events, and this connection was furthered by Lamarck in his *Annuaires Météorologiques*, much to the despair of his scientific peers and, more important, to the disdain of the emperor.

The need for meteorology to be separated from its zodiacal context had been voiced on and off since the time of Aristotle, and Napoleon, who shared the growing scientific impatience with Lamarck, had been looking for an opportunity to encourage him away from his astrometeorological efforts. The opportunity was granted one winter's afternoon in Paris in 1809, when Lamarck attempted to make a formal public present to Napoleon of a bound copy of his latest work, the *Philosophie Zoologique*. The emperor impatiently refused the present, mistakenly believing it to be the latest installment of the detested *Annuaires*. Taking the opportunity, there and then, he offered a lengthy and stinging rebuke to Lamarck for his hapless contributions to the science of the atmosphere, instructing him to forget all about the moon and the clouds and his astrometeorological musings, and to concentrate instead on evolutionary biology, the science of the future in France.

Napoleon, who had only recently abolished the use of the French revolutionary calendar, intended the same fate for all associated almanacs and *Annuaires*. The emperor's sincerity in the matter was obvious, and Lamarck, confused, humiliated, and utterly at a loss for words, saw his meteorological career come to a sad and sudden end. He was never to write about the subject again, apart from an entry in the *Nouveau Dictionnaire d'Histoire Naturelle*, in which he complained about his treatment from Napoleon.[22] After a life marked by speculative toil and administrative quarreling, he died blind, penniless, and virtually unknown in 1829.

Ironically, his intended gift, the *Philosophie Zoologique*, would have pleased the emperor enormously, treating as it did some of the key debates in biological science, which Darwin was later to address with greater authority and range. But Lamarck's tragic destiny was not to be deflected from its course. His star fell rapidly from public view, and as it did so, the field of nephology was definitively cleared for the language of citizen Howard.

Chapter Seven

Publication

Oft, as he travers'd the cerulean field,
And mark'd the Clouds that drove before the wind;
Ten thousand glorious systems would he build,
Ten thousand great ideas fill'd his mind;
But with the clouds they fled, and left no trace behind.

James Thomson, 1748[1]

When we think of the epoch-making developments that emerged from the revolutions in science and technology, we tend to think of the canal system, the steam engine, or Richard Arkwright's incomparable water frame. Yet one of the greatest—and one of the most overlooked—developments of all was the humble scientific periodical. The rise of the periodical was one of the major scientific advances of all time, for without the periodical there can be little scientific or technological communication, and without communication there can be no community. Knowledge means nothing, after all, if it is not widely shared. "Only everyone can know the truth," as Goethe once brilliantly put it. Little wonder that the Internet, with its phenomenal powers of transmission and recall, has been hailed as the harbinger of another, newer kind of industrial revolution. Like the Internet, periodicals had the power of immediacy, yet bound into annual numbers, they also built up to form the chronicle of a collective intellectual journey. Journals had the advantage over books in a number of ways: they were cheaper to buy, they were more easily transportable, and they proved better at the job of concisely communicating facts and recent events. "The Post-office has done as much to foster science as the British Association itself," as a contributor to the *Quarterly Journal* of the Meteorological Society once pointed out, and the impact of the fast and cheap dispatch of letters, books, and periodicals throughout the literate world should never be underestimated.[2]

The most successful early example of the science-magazine formula was the *Philosophical Transactions* of the Royal Society, but they had suffered from expense and a limited circulation comprised of Royal Society members and their associates, many of whom were "research inactive" aristocratic patrons. A growing need was felt for the scientific journal to be updated to a modern format, one that was aimed at a wider and more appreciative audience and, moreover, one that did not claim copyright on all its submitted material. Something new was needed to reflect the growing place of the sciences in public life, and someone new was needed to promote it.

As an energetic, ambitious, and already successful publisher, Alexander Tilloch was just the man for the job.

Alexander Tilloch had been born in Glasgow in 1759, the son of John Tulloch, a successful tobacco merchant and city magistrate. Like Howard, he came from a newly wealthy family, although unlike Howard, his youthful circumstances had been both comfortable and intellectually agreeable. Following an education at Glasgow University, at that time one of the world's greatest academies, the 28-year-old Alexander Tulloch moved to London in 1787. He took the opportunity as he did so to anglify his surname to the easier-sounding Tilloch. Perhaps he had been made to feel, like Boswell before him, the hindrance of a Scottish upbringing in eighteenth-century England.

With his father's money, Tulloch, now Tilloch, bought controlling shares in *The Star and Evening Advertiser* and promptly set about transforming the year-old newspaper into a vehicle for his own ambitions. Tilloch's gifts as a periodical publisher lay in his eye for the tastes of the moment, and under his leadership the *Star* soon gained a modish reputation for itself. When he persuaded Robert Burns, "the ploughing poet," to contribute some libelous political squibs during the spring of 1789, the scoop led to an agreeable and lucrative controversy. Burns went on publicly to deny authorship of the poems, one of which lampooned his patron the Duchess of Gordon ("Her Grace was mucklest of them aw"), but his protests cut little ice in London circles.[3] There was no doubting Tilloch and his team's considerable gift for timing and publicity, as well as their genius for the popular touch. Burns, then at the height of his newly found fame, was for Tilloch's London readers an exotic and appealing catch: the rhyming farmhand from the Celtic periphery, whose irreverence only added to his charm.

Tilloch soon recognized, however, that other, more lasting tastes were coming into the ascendant. Science and technology, in all their varieties of forms and applications, were occupying more and more of his readers' attention. It seemed to him that the time had clearly come to start adding to the stable, and when Tilloch founded the *Philosophical Magazine* in 1798, he was among the first to have recognized the growing publishing possibilities afforded by the rise of popular science. It was apparent to him that the audiences then clamoring to attend lectures and demonstrations could easily be transformed into a readership. He had firsthand experience of the taste for science, having joined the Askesian Society soon after its founding. In his capacity as the *Philosophical Magazine*'s roving owner-editor, he was always on the lookout for promising material. And on that December evening in 1802, he knew that he had found it. He made sure to sign up Luke Howard's essay on clouds before anyone else could do the same.

In fact, Tilloch's magazine had only one serious rival: William Nicholson's *Journal of Natural Philosophy, Chemistry and the Arts*, known to its coterie of readers as

Nicholson's. Nicholson's journal had been founded a year before Tilloch's magazine, but it had soon begun to feel the effects of the rise of the newer publication. Tilloch's had outstripped Nicholson's reputation and, more importantly in those entrepreneurial times, it had stolen much of its circulation. Nicholson was a failed schoolmaster who was fated to live his entire life in financially difficult circumstances. Even his famous journal couldn't save him, especially now that it had been put at such a disadvantage by Tilloch's. Nicholson's was really only a forum for reviews, offering discussions of recent publications rather than a direct and exciting engagement with scientific work in progress. It was this kind of direct engagement that had been the founding principle of Tilloch's publication, and it was this that would ensure its long-running critical and commercial success.

Founded as a general scientific monthly, the *Philosophical Magazine*'s grand purpose was, as announced in the preface to its first collected volume, to "diffuse Philosophical Knowledge among every Class of Society, and to give the Public as early an Account as possible of every thing new or curious in the scientific World." As if to allay any doubts of its extensive brief, the first number, dated June 1798, carried in its 112 octavo pages articles ranging from "an account of two singular Meteors lately seen in France" to the story of a pair of elephants removed from Paris to The Hague, via conjectures on the effects of magnetism on timepieces, hints on testing the purity of wine, experiments on a new formula for fireproof paper, and an account of the Dutch embassy to the court of Peking.

The magazine also contained digests of the meetings of learned institutions on the continent, such as the National Institute in Rome and the reestablished (although no longer Royal) Académie des Sciences in Paris. Most of the readership was at a wide social remove from the culture of foreign institutions, and they relied entirely on such firsthand information as journals such as Tilloch's or Nicholson's could supply. The two rival publications fought it out for a decade, with Nicholson relaunching his *Journal* in a new format in 1802, but the Scotsman continued to maintain the edge over his rapidly despairing rival.

The campaigns of the two editors were fought on a personal level, or at least, on a level of close personal identification with their products. In an unusual step for the times, both Nicholson and Tilloch placed their own names boldly on the title page of every issue. Nicholson was moved to defend the practice by releasing an advertisement in which he claimed that the conventional anonymity prevented "men of reputation from corresponding openly with the Managers of our Periodical Publications," meaning, of course, that he wanted all the world to know his name. The removal of this obstacle, he insisted, could only further the promotion of knowledge. It certainly furthered the promotion of the respective editors, who were soon forced into rejecting accusations of their "vain pretence to superiority."

Nicholson also had cause to defend his use of the first-person singular, the journalistic "I," which he had introduced, against all scientific and scholarly precedent, in papers written by himself. He saw no need to "conceal himself," as he put it, and neither did his watchful rival Alexander Tilloch, who was soon to prove himself a better player of all such games of publishing personality.

Media proprietorship and editorship were becoming new routes to public recognition, a scenario that, of course, remains familiar today as the background to wars of circulation. Nicholson adopted new editorial strategies designed to keep his title afloat, even introducing lifestyle features such as essays on the art of shaving and other "Philosophical Disquisitions on the Processes of common Life"; but by 1813 the game was well and truly up. Tilloch bought out the rival production, collapsing it into his already expanded magazine, which he then renamed, in an inclusive gesture, the *Philosophical Magazine and Journal.* The fight with Nicholson was over, even though by 1802 the *Philosophical Magazine* had long been the undisputed leader of its field, with one of the most loyal readerships around.

Tilloch's readers were not members of the old landed cultural elite, whose timeworn schedule—from Eton to Oxford to a comfortable seat in the House of Commons and then, perhaps, a fellowship of the Royal Society—was beginning to outlive its own momentum. The readers of the *Philosophical Magazine* were, by contrast, hardworking members of the professional, practical classes who were beginning to swell the audiences of the fast-proliferating lectures, demonstrations, and museum exhibitions. These were the men and women who would go on to join the burgeoning literary and philosophical institutions, the free public libraries, and then, in later decades, the new universities, achieving through education the characteristically modern advantage of intellectual capital.

This readership, broader than that of the *Philosophical Transactions*, was also in many respects better informed, at least on day-to-day technical matters, and it was certainly more demanding. It included artisans and industrialists, professional and working people, some of whom were wealthy and successful, most of whom were not. Learning was for them a new and uplifting opportunity, not one of the languid privileges of a fortunate birth. This was reflected in the tone and style of the publication, which sought at all costs to avoid the narrative culs-de-sac of theoretical ennui. From the first issue, for example, Tilloch employed the practice of dividing major articles into monthly installments, each part ending with the tantalizing and effective enjoinder: *to be continued.* Subscriptions to the publication soared.

Members of the Plough Court Askesian Society, like William Allen and his friend and partner Luke Howard, were typical readers and subscribers, and as Tilloch had already discovered to his profit, they also made excellent contributors. By December 1802, when Luke Howard was invited to contribute his essay on clouds, the *Philosophical Magazine* was easily the best-known scientific publication in Britain. Howard would therefore have had every reason to be as excited as he was by the publishing prospect ahead of him.

Howard was growing ever more convinced that his ideas about cloud shapes were sound. They had, after all, made a well-received lecture at the Askesian Society meeting. But would they hold up in the cold light of print? Everybody he spoke to on the matter seemed to think so. William Allen and the other Askesians were characteristically supportive, and the young Silvanus Bevan of the Plough Court pharmacy (one of the great-grandsons of the founder), who had spent his evenings helping Howard prepare the illustrations, was happy to think that his name, too, was soon to appear in print in an acknowledgment at the end. Things seemed to be happening at an extraordinary pace, as they always did when Tilloch was around. Suddenly a new chapter in the history of clouds was about to be written and published.

Howard's editorial meetings with Tilloch were brief but encouraging, although in truth, they were not meetings so much as a series of spoken instructions:

"Rewrite it, expand it, rearrange the sections, have the drawings clarified and engraved, then send it on to me as soon as you can.—What might you be waiting for?"

Howard went away as instructed and, through Christmas and the New Year, did more, and more wide-ranging, work on the paper. He added many earlier observations that, for the sake of brevity, he had left out of his lecture, and he continued along new and less certain lines of thought, some of which had been suggested to him during long discussions that had followed on from the lecture of the previous year. He added sections on the formation of dew and on the evaporation of water at different temperatures, supplementing them with lengthy citations from an essay written by his friend John Dalton, a fellow Quaker, who taught natural philosophy at the newly founded Dissenters' College in Manchester.

John Dalton was a key personality in the scientific scene of the northwest corner of England. As varied in his interests as he was in his contacts, he worked on color blindness (a condition from which he suffered and of which, amazingly, he made a self-diagnosis), on the solubility of gases, and on the atomic constitution of matter, as well as on meteorology, the earliest and longest-lived of his scientific pursuits. Chronically short of money, he built his own instruments from parts he collected here and there, using them to compile a weather journal that he kept for well over half a century.

Dalton's journals formed the basis of his first book, the *Meteorological Essays* of 1793, in which he began to make progress from the mere accumulation of weather reports to an attempted analysis of the physics of the atmosphere. It was Dalton who first examined the relationship between condensation and the expansion of air, and who first realized that cloud droplets continually fall, rather than float in the air as they appear to. Once water has condensed from its vaporous state into liquid droplets, it is at the mercy of gravitational pull. Convection provides the upward momentum, while gravity provides the reverse.

This idea, as pursued by Dalton, that clouds obeyed the simple laws of physics was an inspiring one for Howard, who continued all the while with his rewrite. He was particularly keen to expand the sections that dealt with the physics of the formation of clouds, which he attempted to call "nubification." This umbrella term referred to the entire progress of vapor, from its uptake from the surface of the earth to "the production of a cloud consisting of visible drops, and confined to a certain space in the atmosphere."[4] "Nubification" covered the entire spectrum of nephological interest, but unlike Howard's other terms, it was never really likely to catch on.

Howard's wife, Mariabella, meanwhile, gave birth to another child on January 26, 1803, and attending to her and the baby daughter, whom they named Elizabeth after his mother and sister, gave him some respite from his nephological labors.

As a parent Luke Howard was more openly loving and understanding than his own father, Robert, had been. His children and grandchildren were to grow up to treasure his gentle, instructive company, even though, according to one of them, he "seemed always to be thinking of something very far away."[5] Indeed, throughout the birth of his daughter Elizabeth he had been as cloud-haunted as ever, working up the results of his lifelong observations during the quieter phases of the confinement. By the time he sent the rewritten draft to Alexander Tilloch in April 1803, the essay had grown from its origins as a few pages of handwritten notes into a finished article of nearly fifteen thousand words, spread over fifty quarto sheets. He had never written anything of such quantity or quality before. Despite himself, Howard was proud of his efforts, for it was clearly a major piece of work, containing not merely the seeds of a new idea but a fully evolved breakthrough in thought.

Clouds, far from being the mere "airy nothings" of the landscape, had been made the subjects of "grave theory and practical research . . . now shewn to be governed, in their production, suspension and destruction, by the same fixed Laws which pervade every other department of Nature."[6] Tilloch, who despite his tone was a serious and considerate editor, knew that the essay would be worth publishing in its entirety. A bit of tidying up, a few minor amendments, and it would soon be ready for the press.

Alexander Tilloch had scooped again.

At the heart of Luke Howard's essay, *On the Modifications of Clouds,* lay the penetrating—and Lamarckian—insight that clouds have many individual shapes but few basic forms. Both shapes and forms are due to physical processes that affect water present in the atmosphere, whether in its gaseous, liquid, or solid form. Although unexpected complexities and complications soon arise because of the instability of the circulating atmosphere, the physical principles of cloud formation are as easily understood as any other natural process. Clouds were no longer exempt from human comprehension, and Howard, in contributing both a system of analysis and a full Latin nomenclature covering their families and genera, had contributed more than anyone to easing the path of understanding.[7]

Cloud formation is dependent on the temperature, humidity, and pressure of air in which water vapor is present. The warmer the air pocket, the more evaporated water it can hold unseen. Conversely, as air cools it can hold less and less water vapor until it reaches what is known as the dewpoint: the temperature at which water vapor condenses into visible liquid droplets. Once air has reached its dewpoint, it is said to be saturated, with relative humidity at 100 percent. Clouds form when air has risen through convection (or any other form of lifting) and has then cooled at the dewpoint and condensed around some of the particles naturally present in the atmosphere, known as condensation nuclei. These ever present microscopic particles of sea salt, pollen grains, dust, or smoke from fires and volcanoes, as well as gaseous exhalations from plants, play an important part in determining the patterns of climate and weather. As has already been seen, the circling veil of dust can affect long-term climate with dramatic, and sometimes devastating, results; yet its function in local cloud formation is vital. In fact, without particles or impurities to cling to, absolutely clean air has great difficulty condensing its vapor and will supersaturate instead, holding more water vapor than might seem possible until suitable nuclei are found.

Once condensation has occurred, the tiny individual water droplets that form around the specks are small enough to remain suspended in the atmosphere, where, bundled together in the billions, they make up a visible cloud. Individually, these cloud droplets measure a mere millionth of a millimeter across. Depending on the temperature of the air in which it forms, a cloud may also be made up of billions of miniature solid ice crystals as well as the liquid droplets. Very cold upper air will freeze water molecules into ice, forming the distinctive, high-level, fibrous clouds named cirrus by Howard in his lecture. When gravity begins to pull the crystals downward, and a wind blows across the slowly falling cirrus, it can be streaked into its familiar mare's-tail formation, known as cirrus uncinus (*uncinus* from the Latin word for "hooked").

On a warm, sunny day, the water in a low-level cloud such as a fair-weather cumulus will continue to evaporate upward into vapor while being replenished from the surface of the earth below. This gives the cloud an aspect of stability while in fact it undergoes a constant and unstable process of change, as is dramatically visible in speeded-up film. Even the stillest cloud is in a restless turmoil of vaporous exchange. Latent heat is released by the vapor as it condenses, and this in turn helps keep lifted parcels of air buoyant and unstable. Depending on the force of the wind in the upper air, fast-changing cumulus clouds may seem either to hang still or to wander across the sky like lonely Wordsworthian fragments. However fast or slowly they move, their average life expectancy is a mere ten minutes.

Little fluffy cumulus clouds are produced by small-scale local convection. If conditions allow it, however, cumulus clouds can grow massive through replenishment by large-scale systems of warm rising air, which as they cool add more and more condensing moisture to the cloud. As the cloud continues to grow upward through convection, the latent heat given off adds further buoyancy and instability to the structure. If it rises further, ice crystals will begin to form at the top from supercooled water droplets. Their positive electrical charge, interacting with the negative charge building at the liquid base of the cloud, is what leads to lightning, a phenomenon produced only within the larger cumulonimbus structures. Most lightning stays within the cloud, which is what we glimpse as sheet lightning, but some, of course, makes its way to the ground, with sometimes devastating results.

Luke Howard was the first meteorologist to systematically apply recent electrical theories in an attempt to explain everyday atmospheric events, and though he may have overemphasized the role played by electricity in the formation and life span of clouds, he nevertheless instituted an area of nephology that is still being debated today. It may well be, for example, that it is the sudden redistribution of electrical charge caused by lightning within a cloud that helps cause the onset of rain, as Howard tentatively suggested. And shifts in electrical state almost certainly bring about coalescence, the coming together of liquid droplets, which is a known prerequisite of rainfall.

But electrical theory aside, Howard's main contribution to atmospheric science remains the naming of the clouds in his lecture of 1802. When the lecture appeared in print the following summer, his breakthrough was prefaced with the following words:

> In order to enable the meteorologist to apply the key of analysis to the experience of others, as well as to record his own with brevity and precision, it may perhaps be allowable to introduce a methodological nomenclature, applicable to the various forms of suspended water, or, in other words, to the modifications of cloud.[8]

Howard referred to *modifications* of clouds rather than to genera or species because he wanted most of all to emphasize that the particular form assumed by any one cloud was likely to vary at any given moment, due to the shifting instability of the atmosphere. Although the Linnaean system of classification lent Howard his organizational model, its terms when applied to clouds gave the wrong impression of fixity. It was the very mutability of clouds, their very lack of fixity, that offered the particular challenge to their classification. As Lamarck had recently discovered to his cost, clouds present a moving target, so any scientific account of them would have to be true to a dynamic rather than a fixed view of the world. As an evolutionist (or, more exactly, as a "transformist"), Lamarck shared the view that nature was a fundamentally changeable yet fundamentally ordered system, declaring that "nothing in nature is immutable; all objects are subject to continual and inevitable changes which arise from the essential order of things."[9] He recognized, as did Howard, that a simple basic cloud structure was needed from the outset, but his failure lay in the delivery of a satisfactory model. What he lacked most of all was organizational insight, and the loose compilation of textures and shapes that he proposed in its stead was overly complex and unclear.

By contrast, Luke Howard's claim that "there are three simple and distinct modifications, in any one of which the aggregate of minute drops called a cloud may be formed, increase to its greatest extent, and finally decrease and disappear" was a bold intervention, made even bolder when followed up by the most significant part of the argument—the breakthrough: "but the same aggregate which has been formed in one modification, upon a change in the attendant circumstances, may pass into another."[10] Clouds could change both generically and specifically by effecting transitions between their forms. They could pass, one into the other, not only between individual modifications but between entire families of forms. Cumuliform clouds might spread themselves out to become stratiform clouds; then again, they might evaporate and rise through convection to form higher, cirriform structures. The shapes made by vapor merge and demerge—rising through convection, falling through gravity—yet nephology can chart every stage of their progress. This one penetrating clause laid the main plank for the foundation of the science of clouds. Clouds could now be read more clearly, as the visible signs of vast atmospheric processes. Altogether, the breakthrough was breathtaking: it granted the clouds their mobility instead of willing them to be still for the benefit of science.

The age was already besotted with taxonomy, and Howard's new typology of suspended water struck his readers as a bold intervention. So much had been named and classified in the previous fifty years, so much had been Latinized for study or stuffing, from the greatest of the animals, plants, and fishes all the way down to the microbes, minerals, and the first discoveries among the chemical elements. The eighteenth century had been the great age of naming and fixing, with Linnaeus and Samuel Johnson enthroned as its ruling and rule-giving lexicographers. Both the language of nature and the nature of language were being drawn up and standardized for the benefit of compatible usage.

But the naming of clouds was a different kind of gesture for the hand of classification to have made. Here was the naming not of a solid, stable thing but of a series of self-canceling evanescences. Here was the naming of a fugitive presence that hastened to its onward dissolution. Here was the naming of *clouds*.

This was how the first audiences would have read of the advance in the pages of the *Philosophical Magazine*:

> *On the Modifications of Clouds, and on the Principles of their Production, Suspension, and Destruction; being the Substance of an Essay read before the Askesian Society in the Session 1802-3. By* Luke Howard, *Esq.*
>
> The simple modifications are thus named and defined:
>
> 1. Cirrus. *Def. Nubes cirrata, tenuissima, quæ undique crescat.*
> Parallel, flexuous, or diverging fibres, extensible in any or in all directions.
>
> 2. Cumulus. *Def. Nubes cumulata, densa, sursum crescens.*
> Convex or conical heaps, increasing upward from a horizontal base.

3. Stratus. *Def. Nubes strata, aquæ modo expansa, deorsum crescens.*
A widely extended, continuous, horizontal sheet, increasing from below upward. This application of the Latin word *stratus* is a little forced. But the substantive, *stratum*, did not agree in its termination with the other two.

The intermediate Modifications which require to be noticed are:

4. Cirro-cumulus. *Def. Nubeculæ densiores subrotundæ et quasi in agmine appositæ.*
Small, well defined roundish masses, in close horizontal arrangement or contact.

5. Cirro-stratus. *Def. Nubes extenuata sub-concava vel undulata. Nubeculæ hujus modi appositæ.*
Horizontal or slightly inclined masses attenuated towards a part or the whole of their circumference, bent downward, or undulated, separate, or in groups consisting of small clouds having these characters.

The compound modifications are:

6. Cumulo-stratus. *Def. Nubes densa, basim planam undique supercrescens, vel cujus moles longinqua videtur partim plana partim cumulata.*
The Cirro-stratus blended with the Cumulus, and either appearing intermixed with the heaps of the latter or superadding a wide-spread structure to its base.

7. Cumulo-cirro-stratus *vel* Nimbus. *Def. Nubes vel nubium congeries* [superné cirrata] *pluviam effundens.*
The rain cloud. A cloud, or system of clouds from which rain is falling. It is a horizontal sheet, above which the Cirrus spreads, while the Cumulus enters it laterally and from beneath.[11]

These collective opening definitions, soon to be widely circulated and reprinted, served to introduce both the classification and the nomenclature. They operated as a summary and as a key for readers to refer to, as the essay developed from there into a series of meditative entries on the seven individual cloud genera, each marshaled under the heading of its newly coined term:

Of the Cirrus.

Clouds in this modification appear to have the least density, the greatest elevation, and the greatest variety of extent and direction. They are the earliest appearance after serene weather. They are first indicated by a few threads pencilled, as it were, on the sky. These increase in length, and new ones are in the mean time added laterally. Often the first-formed threads serve as stems to support numerous branches, which in their turn give rise to others.

The increase is sometimes perfectly indeterminate, at others it has a very decided direction. Thus the first few threads being once formed, the remainder shall be propagated either in one, two, or more directions laterally, or obliquely upward or downward, the direction being often the same in a great number of clouds visible at the same time: for the oblique descending tufts shall appear to converge towards a point on the horizon, and the long straight streaks to meet in opposite points therein; which is the optical effect of parallel extension.

Their duration is uncertain, varying from a few minutes after the first appearance to an extent of many hours. It is long when they appear alone and at great heights, and shorter when they are formed lower and in the vicinity of other clouds.

This modification, although in appearance almost motionless, is intimately connected with the variable motions of the atmosphere. Considering that clouds of this kind have long been deemed a prognostic of wind, it is extraordinary that the nature of this connection should not have been more studied, as the knowledge of it might have been productive of useful results.

In fair weather, with light variable breezes, the sky is seldom quite clear of small groups of the oblique cirrus, which frequently come on from the leeward, and the direction of their increase is to windward. Continued wet weather is attended with horizontal sheets of this cloud, which subside quickly and pass to the cirro-stratus.

Cirrus

Before storms they appear lower and denser, and usually in the quarter opposite to that from which the storm arises. Steady high winds are also preceded and attended by streaks running quite across the sky in the direction they blow in.

Of the Cumulus.

Clouds in this modification are commonly of the most dense structure: they are formed in the lower atmosphere, and move along with the current which is next the earth.

A small irregular spot first appears, and is, as it were, the nucleus on which they increase. The lower surface continues irregularly plane, while the upper rises into conical or hemispherical heaps; which may afterwards continue long nearly of the same bulk, or rapidly rise to mountains.

In the former case they are usually numerous and near together, in the latter few and distant; but whether there are few or many, their bases always lie nearly in one horizontal plane, and their increase upward is somewhat proportionate to the extent of the base, and nearly alike in many that appear at once.

Their appearance, increase, and disappearance, in fair weather, are often periodical, and keep pace with the temperature of the day. Thus they will begin to form some hours after sun-rise, arrive at their maximum in the hottest part of the afternoon, then go on diminishing and totally disperse about sun-set.

But in changeable weather they partake of the vicissitudes of the atmosphere; sometimes evaporating almost as soon as formed, at others suddenly forming and as quickly passing to the compound modifications.

The cumulus of fair weather has a moderate elevation and extent, and a well defined rounded surface. Previous to rain it increases more rapidly, appears lower in the atmosphere, and with its surface full of loose fleeces or protuberances.

The formation of large cumuli to leeward in a strong wind, indicates the approach of a calm with rain. When they do not disappear or subside about sun-set, but continue to rise, thunder is to be expected in the night.

Cumulus

Independently of the beauty and magnificence it adds to the face of nature, the cumulus serves to skreen the earth from the direct rays of the sun, by its multiplied reflections to diffuse, and, as it were, economise the light, and also to convey the product of evaporation to a distance from the place of its origin. The relations of the cumulus with the state of the barometer, &c. have not yet been enough attended to.

Of the Stratus.

This modification has a mean degree of density.

It is the lowest of the clouds, since its inferior surface commonly rests on the earth or water.

Contrary to the last, which may be considered as belonging to the day, this is properly the cloud of night; the time of its first appearance being about sun-set. It comprehends all those creeping mists which in calm evening ascend in spreading sheets (like an inundation of water) from the bottom of valleys and the surface of lakes, rivers, &c.

Stratus

Its duration is frequently through the night.

On the return of the sun the level surface of this cloud begins to put on the appearance of cumulus, the whole at the same time separating from the ground. The continuity is next destroyed, and the cloud ascends and evaporates, or passes off with the appearance of the nascent cumulus.

This has been long experienced as a prognostic of fair weather, and indeed there is none more serene than that which is ushered in by it. The relation of the stratus to the state of the atmosphere as indicated by the barometer, &c. appears notwithstanding to have passed hitherto without due attention.

Of the Cirro-cumulus.

The cirrus having continued for some time increasing or stationary, usually passes either to the cirro-cumulus or the cirro-stratus, at the same time descending to a lower station in the atmosphere.

The cirro-cumulus is formed from a cirrus, or from a number of small separate cirri, by the fibres collapsing, as it were, and passing into small roundish masses, in which the texture of the cirrus is no longer discernible, although they still retain somewhat of the same relative arrangement. This change takes place either throughout the whole mass at once, or progressively from one extremity to the other. In either case, the same effect is produced on a number of adjacent cirri at the same time and in the same order. It appears in some instances to be accelerated by the approach of other clouds.

Cirro-cumulus

This modification forms a very beautiful sky, sometimes exhibiting numerous distinct beds of these small connected clouds, floating at different altitudes.

The cirro-cumulus is frequent in summer, and is attendant on warm and dry weather. It is also occasionally and more sparingly seen in the intervals of showers, and in winter. It may either evaporate, or pass to the cirrus or cirro-stratus.

Of the Cirro-stratus.

This cloud appears to result from the subsidence of the fibres of the cirrus to a horizontal position, at the same time that they approach towards each other laterally. The form and relative position, when seen in the distance, frequently give the idea of shoals of fish. Yet in this, as in other instances, the structure must be attended to rather than the form, which varies much, presenting at other times the appearance of parallel bars, interwoven streaks like the grain of polished wood, &c. It is always thickest in the middle, or at one extremity, and extenuated towards the edge. The distinct appearance of a cirrus does not always precede the production of this and the last modification.

Cirro-stratus

The cirro-stratus precedes wind and rain, the near or distant approach of which may sometimes be estimated from its greater or less abundance and permanence. It is almost always to be seen in the intervals of storms. Sometimes this and the cirro-cumulus appear together in the sky, and even alternate with each other in the same cloud, when the different evolutions which ensue are a curious spectacle, and a judgement may be formed of the weather likely to ensue by observing which modification prevails at last. The cirro-stratus is the modification which most frequently and completely exhibits the phænomena of the solar and lunar halo, and (as supposed from a few observations) the parhelion and paraselene also. Hence the reason of the prognostic for foul weather, commonly drawn from the appearance of halo.

This modification is on this account more peculiarly worthy of investigation. Little is yet ascertained of the relations of this and the last modification with the barometer, &c. although, as may be readily supposed, they have been found to accord with opposite indications of those instruments.

Of the Cumulo-stratus.

The different modifications which have just been treated of sometimes give place to each other, at other times two or more appear in the same sky; but in this case the clouds in the same modification lie mostly in the same plane of elevation, those which are more elevated appearing through the intervals of the lower, or the latter showing dark against the lighter ones above them. When the cumulus increases rapidly, a cirro-stratus is frequently seen to form around its summit, reposing thereon as on a mountain, while the former cloud continues discernible in some degree through it. This state continues but a short time. The cirro-stratus speedily becomes denser and spreads, while the superior part of the cumulus extends itself and passes into it, the base continuing as before, and the convex protuberances changing their position till they present themselves laterally and downward. More rarely the cumulus alone performs this evolution, and its superior part constitutes the incumbent cirro-stratus.

Cumulo-stratus

In either case a large lofty dense cloud is formed,

which may be compared to a mushroom with a very thick short stem. But when a whole sky is crowded with this modification, the appearances are more indistinct. The cumulus rises through the interstices of the superior clouds, and the whole, seen as it passes off in the distant horizon, presents to the fancy mountains covered with snow, intersected with darker ridges and lakes of water, rocks and towers, &c. The distinct cumulo-stratus is formed in the interval between the first appearance of the fleecy cumulus and the commencement of rain, while the lower atmosphere is yet too dry; also during the approach of thunder storms: the indistinct appearance of it is chiefly in the longer or shorter intervals of showers of rain, snow, or hail.

The cumulo-stratus chiefly affects a mean state of the atmosphere as to pressure and temperature; but in this respect, like the other modifications, it affords much room for future observation.

Of the Nimbus, or Cumulo-cirro-stratus.

Clouds in any one of the preceding modifications, at the same degree of elevation, or in two or more of them at different elevations, may increase so as completely to obscure the sky, and at times put on an appearance of density which to the inexperienced observer indicates the speedy commencement of rain. It is nevertheless extremely probable, as well from attentive observation as from a consideration of the several modes of their production, that the clouds while in any one of these states do not at any time let fall rain.

Before this effect takes place they have been uniformly found to undergo a change, attended with appearances sufficiently remarkable to constitute a distinct modification. These appearances, when the rain happens over our heads, are but imperfectly seen. We can then only observe, before the arrival of the denser and lower clouds, or through their interstices, that there exists, *at a greater altitude* a thin light veil, or at least a hazy turbidness. When this has considerably increased we see the lower clouds spread themselves till they unite in all points and form one uniform sheet. The rain then commences, and the lower clouds, arriving from the windward, move under this sheet and are successively lost in it. When the latter cease to arrive, or when the sheet breaks, every one's experience teaches him to expect an abatement or cessation of rain.

But there often follows, what seems hitherto to have been un-

Nimbus

noticed, an immediate and great addition to the quantity of cloud. At the same time the actual *obscurity* is lessened, because the arrangement, which now returns, gives freer passage to the rays of light: for on the cessation of rain the lower broken clouds which remain rise into cumuli, and the superior sheet puts on the various forms of the cirro-stratus, sometimes passing to the cirro-cumulus . . . The nimbus, although in itself one of the least beautiful clouds, is yet now and then superbly decorated by its attendant the rainbow; which can only be seen in perfection when backed by the widely extended uniform gloom of this modification.[12]

From the rain cloud, Howard moved on naturally to a discussion of the formation and precipitation of rain itself. This was a subject to which he gave a great deal of subsequent attention, and he refined his later definition of "nimbus" to accommodate his further thoughts on rain, conceding that the formation of two distinct layers "is not the only mode in which rain may be produced out of the clouds."[13] He was right. Most rain, especially persistent rain or drizzle, falls from bands of stratiform clouds, while brief summer showers fall from the more isolated cumuliform clouds.

Precipitation tends to form in the colder sections of the cloud, where the temperature is at or below freezing point, and where there is a mixture of ice crystals and tiny, supercooled water droplets. These unstable droplets freeze when they come into contact with ice particles and then begin to grow as crystals rather than as liquid drops. Such crystals will start to fall through the cloud once they are heavy enough, continuing to grow through coalescence with any liquid droplets or miniature ice crystals that they encounter on the way down. As ice builds up on their surfaces, these crystals, growing heavier and falling ever faster, come together at this stage to form snowflakes. Clouds, as Descartes had correctly guessed, are full of snow, and most precipitation begins in the form of snow; it is the temperature of the air through which it subsequently falls that determines the form in which it lands on earth. If the air is cold enough, or if the flakes fall fast enough, the precipitation will land in its original form.

Liquid rain, on the other hand, consists largely of melted snowflakes that have joined up during their descent through warmer air to form drops of varying sizes, from the finest drizzle from sheets of gray stratus (with an 0.2 millimeter drop size) to the heaviest downpours from large cumulonimbus clouds (with drops measuring up to half a centimeter across), massive structures that can hold up to half a million tons of water. Raindrops themselves may merge, break up, and merge again several times before they reach the ground, or may even evaporate entirely. (A shower of rain that never reaches the ground is known as a virga, a twentieth-century term taken from the Latin for "rod.")

Hailstones have a slightly different life story, being formed by warm updrafts of air that throw descending ice pellets back up into the coldest regions of the cloud. The pellets then grow through collision and freezing to create ever bigger and weightier stones. The warmer the unstable updrafts of air, the more journeys the pellets will make back up into the freezing cloud regions, and the bigger the hailstones that will eventually fall unimpeded to the earth. Depending on the convection conditions, on how upwardly mobile the air is, hailstones can grow to an alarming size: the heaviest ever recorded fell in Bangladesh in 1986 during a hailstorm that claimed the lives of ninety-two people. The stone weighed slightly more than a kilogram.

Howard, meanwhile, continued to refine and amend his cloud classification, along with its accompanying descriptions, and in a lecture series given in Tottenham in 1817 (although not published until twenty years later), he reorganized his seven types of cloud using altitude as the ruling consideration. Perhaps by then he had read Lamarck and recognized the value of his tiers:

The Seven Modifications of Cloud are:

1. Cirrus; highest and lightest.
2. Cirrocumulus, } intermediate
3. Cirrostratus, } intermediate
4. Cumulus; detached hemispherical.
5. Cumulostratus; irregular heaped.
6. Nimbus; for Rain.
7. Stratus; mist or fog.[14]

As a Quaker, Luke Howard shared Linnaeus's profoundly religious sense that taxonomy was intended as "a respectful ordering of God's Creation," an outlook reflected by the lack of anything mechanical or life-denying in his classification of clouds.[15] He sought to celebrate rather than to contain nature, and while offering a physical account of the formation and appearance of atmospheric vapor, he allowed nonetheless for the pleasures of reverie and reverence, for a simple delight in the passing of clouds. And therein lay the genius of Luke Howard's approach: by allowing for the mutability of clouds, for their wayward and changeable life over time, it allowed aerial nature to retain the whole of its ancient and sensual appeal in the face of an empirical taxonomy. The physical beauty of the clouds was preserved, as were the dynamic elements of their mystery, undiminished by an otherwise powerful moment of scientific clarity and truth.

Yet Dissenting science also placed a heavy emphasis on practical utility, to which Howard characteristically responded. After predicting, perhaps a touch immodestly for a Quaker, that "the author's modifications will begin to be noted in meteorological registers as they occur, a practice which may be productive of considerable advantage to science," Howard devised a system of shorthand symbols for rapid notation that also remains in use today in a modified version. Designed to "save room and the labour of writing," Howard trusted that "types may be easily formed for printing them. These are advantages not to be despised, when observations are to be noted once or oftener in the day. It is only necessary that they be in-

Cirrus.
Cumulus.
Stratus.
Cirro-cumulus.
Cirro-stratus.
Cumulo-stratus.
Cumulo-cirro-stratus, or Nimbus.

Luke Howard's cloud symbols

serted in a column headed *Clouds.*"[16] For once, the suggestion was immediately taken up, with the *Philosophical Magazine* and others soon running weather charts that featured a cloud column diligently compiled using Howard's recent but fast-ascending terms.

His clouds had arrived and, unlike Lamarck's, they looked as if they were here to stay.

Chapter Eight

Growing Influence

> The tender hues and evanescent forms of the clouds, all afford to persons who know how to view them with a painter's eye, the enjoyment of what I may call a new sense, unknown to those who have not a natural or acquired taste for such studies.
>
> *Charlotte Smith, 1804*[1]

By the end of the summer of 1803, Luke Howard's essay on the modifications of clouds had taken its place in the public domain, where it would go on to be read and discussed wherever new ideas were taken seriously. The essay's appearance was preceded by its reputation, promoted by the members of the Plough Court academy, whose diverse connections throughout the wider scientific world ensured that its contents had already begun to circulate in the months before its actual publication. In contrast to the efforts of the unlucky Lamarck, whose audience for his clouds had featured an unenthused emperor blessed with the power of veto, Luke

Howard's words were sent out to an expectant public readership, an audience ready and waiting for something well out of the ordinary. It was not to be disappointed in its hope.

Given that the means of instant communication by wire still lay half a century in the future, the speed with which these events unfolded is striking even now. To follow the fortunes of Howard's essay both as a written statement and as an autonomous object in the world is to reveal some of the remarkable mechanics of intellectual dissemination at the turn of the nineteenth century. People read, recommended, discussed, passed on, and circulated journals, books, ideas, and objects with the enthusiasm that characterized the period. Important or significant publications were rapidly reissued, reprinted, copied out, or committed to memory by a series of insatiable consumers. And what they were most insatiable for were words: whether written, printed, or spoken aloud, words ruled the pace of culture and of cultural change, and Howard's new language was quickly taken up as part of this wider circulation.

Following its serialization in the July, September, and October editions of the *Philosophical Magazine* in 1803, the essay was published in its own right as an offprint. Howard gave away many of these to his friends and fellow Askesians, who used them on their peri-urban ramblings as guides to the unfolding skies above. Due to its fragility and its fate as a fieldwork companion, it is now an object of extreme rarity, although one, at least, has been preserved in the library of the Friends Meeting House in London, bound into a volume of tracts. It has been unmistakably rained on, having almost certainly been read aloud in the open air during some of the usage that was intended for it by its author.

The thirty-two-page pamphlet (which was reissued several times in the course of the century) went from hand to hand among the scientific and Quaker communities and soon found its way into the offices of the *Annual Review*, via Arthur Aikin of the Askesian Society. The *Annual Review* was a dedicated reviewing publication that had been launched by Aikin in 1802 in an effort to lift some of the growing pressures from the desks of the monthly magazines. Given that the essay was one of the shortest of the almost six hundred new publications reviewed in the volume for 1804, *On the Modifications of Clouds, &c.* was discussed at surprising length, the anonymous reviewer (who was actually John Bostock) justifying the amount of attention given in "a pretty extensive review of this short treatise, because we think it an important work on an important subject."[2]

The essay, according to Bostock, would appeal to anyone "who has viewed the sublime spectacle of a moonlight evening" and been struck by "how much the beauty of the scene has been occasionally heightened by the large round masses of cloud, which not infrequently sail across the firmament, and 'turn forth their silver linings on the night.' "[3] The review, for the most part favorable, verging on the rhapsodic (even quoting Milton for added effect), concluded that "the Method pursued by Mr Howard" was the most likely means yet devised to remove any remaining obstacles to the progress of meteorology.[4] After reproducing the new nomenclature in full, the review went on to discuss the definitions in further detail, expressing one or two reservations, including the somewhat Lamarckian idea that a mere seven families of cloud modifications could not possibly cover the full range of identifiable cloud "species," a term never used by Howard himself. Furthermore, the reviewer cast doubt on the idea that "nimbus" could be regarded as a separate category in its own right, since rain can fall from a number of modifications; this was an issue that Howard would go on to address at a later date.

A particular complaint, however, directed at Howard's choice of language, then received the first of its many airings. This criticism, that Howard had not used "plain English names in a science, the farther improvement of which will, probably, in a considerable degree, depend upon the observations of the unlearned" was a problematic one, and would go on to be taken up elsewhere by Bostock and others, with (as we shall see) potentially disastrous results.[5] Howard, to his irritation, would continually need to defend himself against this charge of obscurantism. In his view, Latin was the natural language of classification, and the *Annual Review* was mistaken in its claims against it, even though, in every other respect, the publication gave an enthusiastic reception to the essay. And whatever else he might have made of the language, it was clear to Bostock, who may well have been a member of the original Plough Court audience, that the key to the study of the clouds had been found. He just didn't like the formal sound of the scientific Latin.

Word of Howard's essay soon spread, and other publications, ranging from general encyclopedias to specialist journals, began to carry their own versions. A digest entitled "Cloud," prepared by Howard himself in 1807, was printed in volume 8 of Abraham Rees's great thirty-nine-volume *Cyclopædia, or Universal Dictionary of Arts, Sciences, and Literature,* to which Howard went on to contribute the entries on rain, dew, Penn, and Quakers.[6] He was by this time fully embarked on the stream of occasional writing that would occupy much of the rest of his working life. He also took the opportunity to oversee the engraving of a new set of cloud illustrations that he had commissioned from the London-based topographical painter Edward Kennion, at a cost (according to his account books) of three pounds and eleven shillings.

Earlier that same year, in January 1807, a new general periodical was launched, edited by the leading Dissenting journalist John Aikin (1747–1822), the father of Arthur Aikin of the *Annual Review* (and Luke Howard's fellow Askesian). An advertisement for the new journal, to be entitled *The Athenæum: A Magazine of Literary and Miscellaneous Information,* outlined its wide-ranging literary, commercial, and scientific contents and promised that, alongside the lists of bankruptcies, corn prices, and "foreign occurrences," a meteorological register would be communicated to the work by "an observer particularly distinguished for his accurate and sagacious remarks upon the phenomena presented by the atmosphere."[7] Luke Howard might have been pleased by the description (he may well of course have written it himself), for he was none other than the subject of the puff.

Howard's appointment to the post of meteorological correspondent was no doubt eased by his connection with Arthur Aikin, the son of the proprietor. Aikin junior was an easygoing young man who had published a well-received geological travel book in 1797 at the age of 24.[8] The intervening decade had seen him transformed into a lecturer on chemistry and chemical manufacturing, the owner of a reviewing journal, a cofounder of the Geological Society, and a major mover and shaker within scientific London. He thus found himself in a position to arrange favors for his friends, most of whom, like him, were non-university-educated scientific activists of energy, wit, and ambition. The little group of Askesian amateurs was beginning to turn professional.

Hired as a meteorological journalist, Howard now had a monthly outlet for his ideas and for his new nomenclature ("a beautiful display of the cirrus cloud all day," reads a typical entry, recorded in August 1807), and he seized the opportunity to encourage the other meteorological correspondents, whose contributions he edited for inclusion in his column, to take it up.[9] One J. S. Stockton, for example, writing from Malton, Yorkshire, in April 1809 could write confidently that "the light and dark Cirro-Stratus were uniformly succeeded by wind and rain, and the Cirro-Cumulus was pretty frequent during the fair intervals," and he sought to pay tribute to the terms that he found himself using.[10] As Howard himself was at pains to point out, his Latin names for the clouds were "but seven in number, and very easy to remember," yet taken together they formed a living language well worth learning to apply.[11]

When the *Athenæum* folded in 1809, because of the financial problems of its proprietor, Howard transferred his meteorological tables to a series of other publications. Most of these reprinted the first half of the original essay to ensure that their readers would be fully equipped to make sense of the expanded weather coverage. William Nicholson's *Journal of Natural Philosophy,* for example (soon to be taken over by Alexander Tilloch), printed a new version of the essay in September 1811, rewritten and retitled "The Natural History of Clouds."[12] In his editorial, Nicholson spoke warmly of the treatise and its author, "long known and valued by the public," and of "that sentiment of obligation, which myself, and the other cultivators of science, must entertain for his researches."[13] He spoke less warmly of the permission fee he had had to pay to Alexander Tilloch.

These tributes carried on for the rest of the decade as journals continued to reprint the essay. The launch issue of Thomas Thomson's *Annals of Philosophy*, for example, carried a digest of Luke Howard's terms, while every monthly issue thereafter carried a meteorological journal compiled by the now prolific Howard, complete with his luminous invocations of the clouds: "the sky about sunset was overspread with Cirrus and Cirrostratus clouds, beautifully tinged with flame colour, red and violet," he wrote, and who could resist the radiance of this heightened, cloud-struck language?[14]

The habit of keeping daily weather observations was never to leave Luke Howard, especially now that it had become so profitable (his father would have been pleased), and a decade later the entire record to that date—Howard's collected journalism, in effect—was issued in book form as *The Climate of London.* The seven-hundred-page book was published in two parts by the Quaker company Phillips of Lombard Street—the first volume appearing in 1818, the second not until 1820. The book confirmed the strength of Howard's growing reputation, and by the time the second volume was in the stores, *The Climate of London* had already been hailed as the founding classic of a pioneering branch of science—urban meteorology—and enthusiastically endorsed by the journals.[15]

The Climate of London was an evocative title, conjuring the idea—novel at the time—that a city might be read by its weather. It was Baudelaire who pointed out, later in the century, that meteorology is an urban pursuit. His cast of flaneurs stroll through the boulevards, measuring their spiritual fortunes against the cinema of drifting clouds, the only city architecture that can be relied on to keep its distance.

The cloud nomenclature featured prominently throughout the text of *The*

Climate of London, with the definitions themselves excerpted as a glossary at the end. When a second, greatly expanded, edition was published in 1833, the entire essay on clouds, by then a well-known article in its own right, was moved to the front to serve as a dramatic opening for the book.

The cloud definitions themselves, meanwhile, continued to turn up everywhere, and, among the other, less tangible manifestations of fame that had begun to color his life, their author began to receive admiring letters addressed to him at the Askesian Society meeting rooms at Plough Court. They were uniformly complimentary, and Howard was embarrassed by the task of acknowledging the praise in his replies. One of these letters, though, was particularly poignant, as its writer had been blind for forty years. John Gough of Kendal had lost his sight during an attack of smallpox as a child, but he could evidently remember the shapes and forms of the clouds that he had seen, especially now that somebody had gone to the trouble of describing and naming them. As he testified in his letter, dated March 30, 1805, he was delighted by the memories they afforded:

> Sir,
> Your observations on the structure and classification of Clouds are very ingenious; in all probability, a person, who can speak more fully on the subject & with greater propriety than myself, will also alow them to be just. In fact I may venture to say that your definitions of

> Cirrus, Cumulus & Stratus, with those of the intermediate Compounds, are so distinct, & probably at the same time so descriptive, that a judicious Meteorologist will hardly neglect to note the recurrence of these appearances in his Journal, by some system of abbreviation; & for any thing I know to the contrary, your scheme of notation is the best that can be devised for the purpose.[16]

Howard replied at length to his "Respected Friend," and, after offering his "hearty acknowledgements" for the flattering letter, he went on to make a revealing observation on his work: "In throwing together my Ideas on Clouds in 1802 I was much embarrassed (as a more profound thinker might be now) in the attempt to reconcile their appearances with the received Principles of Meteorology."[17] This, he said, accounted for what he now regarded as his overreliance on John Dalton's theories of rainfall, which Gough, too, had gone on to question throughout his letter. Howard was glad to find that others were in agreement, and his and Gough's continuing exchange of letters was to be of great help to him in his later amendments to the essay.

He was delighted to have elicited Gough's respect, for Gough himself was a noted botanist, having taught himself the entire Linnaean system by touch. He was also a master of mathematics, zoology, and scoteography—the art of writing in the dark. He might also have become an accomplished musician had his father, a stern

Quaker cast in the mold of Ollive Sims, not stopped him playing the godless violin that an itinerant fiddler had given him. His intellectual accomplishments were to earn him the admiration of his contemporaries, and Coleridge sent his children over the lakes from Keswick to be tutored by "the every way amiable and estimable John Gough of Kendal."[18] Wordsworth, too, was moved to verse by the accomplishments of the blind man of Kendal. "By science led, his genius mounted to the plains of heaven," as his description of Gough in the *Excursion* began:

> . . . the whole countenance alive with thought,
> Fancy and understanding: while the voice
> Discoursed of natural or moral truth
> With eloquence, and such authentic power,
> That, in his presence, humbler knowledge stood
> Abashed, and tender pity overawed.[19]

When the Meteorological Society was formed in 1823, Howard went out of his way to ensure that the very first paper to be read to the group was the one (on the subject of vernal winds) submitted by his estimable friend John Gough.

Yet in spite of these examples of rapid acceptance, the wider reading public was not always immediately willing to take to the new classification. By 1809, for example, the *Gentleman's Magazine* had followed the lead of the *Athenæum* by incorporating the Howard terminology within its own meteorological coverage. But the editors received a letter in 1810 complaining about the use of the new and unfamiliar Latin. The terms should have been printed in English, the anonymous correspondent complained, in an echo of the objection in the *Annual Review*, "because some of the words, being *technical*, are not to be met with in the common dictionaries."[20] Journal readers love a heated correspondence, especially on the subject of the weather, and a flurry of letters on the subject ensued. One of these pointed out that the terms had in fact appeared in Abraham Rees's well-known *Cyclopædia* (although, as the correspondent confidently claimed, "nine out of ten of your Readers do not possess that work").[21]

Struck by this exchange, the *Gentleman's Magazine* decided to run an explanatory guide to the nomenclature of clouds. It appeared in the issue of December 1810 and again in an expanded form in August 1811, complete with a page of plates engraved from drawings supplied by Howard himself, who, having been informed of the controversy, was keen to play his part in clarifying the issue. As the extended article in the *Gentleman's Magazine* suggested, the classification was intended to be used not only by meteorological journalists but also by the reading public, which clearly needed to be reminded from time to time of "the proper technical names" for the clouds.[22]

This need is apparent even today, for in a diary entry written on July 20, 1990, the Conservative Minister Alan Clark wrote of "a towering thunderhead of Alto-Cumulus, precursor of change not just in the weather, but in the Climate."[23] Political metaphors aside, he was in fact applying quite a misnomer in altocumulus, those small fleecy cloudlets of the middle air. He was referring, of course, to cumulonimbus, but he evidently couldn't locate the right term while sitting so late at night at his desk.

And many of Howard's contemporaries felt the same. For them, the reiteration of cloud terms was still not enough to fix them into the memory, the seven "easy" names proving doggedly difficult to remember. His written scientific Latin had, in their view, absolutely no connection with the spoken English language of the day.

It was John Bostock who fired the first shot. Having already raised the issue in the *Annual Review*, he pursued it further by writing to Nicholson's *Journal*, arch-rival of the better-selling *Philosophical Magazine*, to suggest that Howard's "attempted" terminology was "much too confined to be of any great use, and that the hypothesis on which he proceeds is not entirely correct."[24] He then claimed to have come up with a new nomenclature to put in place of Howard's:

Arc: a body of clouds, stretching in nearly parallel lines over a considerable part of the heavens, and converging in a point in the horizon

Linear arc: long parallel lines or threads

Mottled arc: small rounded clouds, lying side by side or in rows

A wreathed arc: resembling a volume of smoke, as it rises from a chimney top

A feathered arc: resembling feathers, having a linear centre and lateral branches

Shaded clouds: when the clouds are formed into rounded masses of greater or lesser extent, one side of which is very much darker than the other side

Piled clouds and *rolling clouds*: large rounded clouds, which appear as if they were heaped and rolled one upon another

Tufts: clouds which resemble bunches of hair, the fibres of which are sometimes disposed in a perfectly irregular manner

Flocks: when clouds form larger and compact masses than those which I have called *tufts*[25]

Bostock conceded that his terms might well be thought to be "very uncouth," but he nevertheless thought that they were better than Howard's, and he waited for the plaudits to flow.

But the retaliation was swift and uncharacteristically severe, for Howard had clearly been angered by what he saw as malice on the part of Bostock. "In common with the whole public, Dr. B. has just the same liberty to reject the nomenclature, which I had to propose it; and from the care with which he avoids using a single term of mine, I perceive he intends to avail himself of this," wrote Howard at the beginning of a comprehensive rebuttal of the list of newly suggested terms. "In the interval since 1804," he went on, "I have not seen sufficient cause to disturb my original plan by adding or suppressing modifications. It is not that I am vain enough to deem it perfect, but I believe it answers the intended purpose."[26]

Bostock's terms, on the other hand, were a lot more than "uncouth"; they were "inaccurate and imperfect," and lacking in precision. The "*mottled* and *wreathed* arc" were clearly "varieties of cirrocumulus," and "*tufts* and *flocks*" were "varieties of the cirrus," but what on earth was a "*rolling* cloud"? "I have not yet detected it; and it seems too poetical, if, as I conjecture, it is so named, because its parts, if solid, would roll when on an inclined plane."[27] But it was for Bostock's abandonment of scientific Latin that Howard reserved his deepest scorn: "Surely the unlearned can learn, as they have done heretofore. *Alphabet*, which is Greek curtailed, is as well understood as *a*, *b*, *c*; *zenith* and *nadir* are Arabic; and as for *Latin*, our Scotch gardeners can talk it fluently"—unlike, this implied, the learned Bostock, who was taken aback by the force of Howard's defense. In a short, hurt letter that appeared in the subsequent issue of Nicholson's *Journal*, he expressed his deep concern that Howard "was offended at me, for rejecting the nomenclature which he proposed," and sought to reassure him and all his readers that it was not done "from any selfish desire 'of making way for my own' in opposition to it."[28] No one had suggested that it was, but the suspicion, introduced by Bostock himself, was enough to settle the issue against him. Nothing more was heard about Dr. Bostock's clouds, and Howard didn't deign to continue the debate.

This would have been the end of the matter, had it not been for the young meteorological correspondent of the *Gentleman's Magazine* who, only a few months later, decided to join the fray. As someone who had already done a great deal to promote Luke Howard's work on clouds, Thomas Forster's intervention came as even more of a shock than had Bostock's:

> When I asked several Artists, who were about to travel over Wales and other mountainous lands, to watch for and sketch the changes of the different forms of the clouds which took place in such places, in order to compare them with those which are common in flat countries, they told me that they could never remember the technical terms, which were made up of Latin or Greek words, which they did not understand; and wished that names could be given to Meteorological

Phænomena, which are formed out of our own tongue. Struck by this remark, I made the following Name-list:

CURL-CLOUD. The old name in Latin by Mr. Howard, is Cirrus, a curl; Cirrulus and curl being the diminutive.

STACKEN-CLOUD, or Cumulus, from the verb to stack, to heap up.

FALL-CLOUD, or Stratus; being the falling, or subsidence of watery particles in the evening.

SONDER CLOUD, or Cirrocumulus, is a sundered cloud, made up of separated orbs. The characteristick of this cloud being the gathering together into a bed, of little clouds, yet so far asunder as not to touch.

WANE-CLOUD, or Cirrostratus; from the waning or subsiding state of this cloud in all its forms.

TWAIN-CLOUD, or Cumulostratus; made often by the twining or uniting of two clouds together.

RAIN-CLOUD, or Nimbus, speaks for itself. So we can have *Storm-cloud, Thunder-cloud, &c.*[29]

For Howard this was even more problematic than before. Forster, the enthusiastic author of these quixotic translations, had been a follower and a friend for a number of years, and had proved effective in his promotion of the set of Latin terms.

Lacking the modesty or reticence of the Quakers (indeed, he was later to convert to Roman Catholicism during a study trip to Italy), the young acolyte had for years been an assiduous campaigner for the clouds according to Howard. He had even sprung to their defense against the attempted incursion of Bostock, and with his own meteorological columns printed in a wide array of journals, including Nicholson's and Tilloch's as well as the *Gentleman's Magazine*, he had played a small but significant part in building up Luke Howard's fame.

But it seems that he had begun to want some of that fame for himself. In 1810 he had started to make amendments to the nomenclature, adding "certain specific names, calculated to express the particular shape, figure, or manner of arrangement" as distinct from the generic modifications themselves. This was only the beginning of the long process of adding cloud species and varieties to Howard's original families and genera, a stage through which every classification needs to go. Howard would have been pleased by these initial supplementary terms, especially as they were in Latin. The terms added by Forster in 1810 and 1811, with his explanations given in parentheses, were as follows:

Comoides ("from its appearing like a distended lock of hair")
Linearis ("straight lines")
Filiformis ("a confused bundle of threads")

Reticularis ("a beautiful network, consisting of light transverse bars or streaks")
Striatus ("composed of long parallel bars")
Undulatus ("finely undulated")
Myoides ("gives the idea of the fibres of muscles")
Planus ("a large continuous sheet")
Petroides ("rocklike and mountainous")
Tuberculatus ("numerous roundish tubercles")
Floccosus ("divided into loose fleeces")[30]

On the whole, this collection of secondary terms had little impact and was largely passed over by the meteorological community, although a number of terms in current usage do resemble them (see Appendix, p. 355). It was in response to this failure, which he blamed on the difficulty of the language, that Thomas Forster started to prepare his translations from Howard's Latin.

Forster had been working on a collected volume of his meteorological work, entitled *Researches about Atmospheric Phænomena,* which was published in 1813. In one sense it had been written in reaction to the recent achievements of Howard, since most of the book was concerned with the classification of atmospheric effects. Forster's aim was to extend the range of the cloud nomenclature to include other kinds of aerial phenomena, such as haloes, mock suns, and coronas, all of which he sought to classify in open emulation of Howard, who was made the subject of the laudatory opening chapter: "Of Mr. Howard's Theory of the Origin and Modifications of Clouds."[31] This chapter, partly a tribute to Howard, partly a reprint of his essay, had been led up to by a preface that was itself a stunning tour de force of imagined meteorological history. In it, Forster traced a line of influence from early man, pictured by the rhapsodic young author as standing in silent awe before the aerial wonders of creation, expressed by the earliest stirrings of the ancient Egyptians and Syrians. The narrative thread was then woven through Aristotle and Theophrastus (who, Forster tells us, "collected all the popular prognostiks of the weather"), through the *Phænomena* of Aratus and the works of Pliny, Virgil, Lucretius, and Seneca, then the long and silent corridor of the Dark Ages, an ignominious vacuum preceding the Renaissance, and on to the era of the luminous moderns, Saussure and de Luc, arriving, finally, at the fountainhead of this winding river of knowledge, where sits the figure of the quiet, godly Englishman Luke Howard, the ornament of his age and the father of modern atmospheric science.[32] It is one of the great scientific tributes of the period, and the bravura of Forster's account captures some of the real excitement of early-nineteenth-century scientific life: popular, celebratory, and up to the minute as it stood upon the threshold of change.

This was all well and good, if a little embarrassing for Howard, but there was still the matter of Forster's English translations to be dealt with. Forster was hoping to see them circulated more widely; since every animal, vegetable, and mineral on earth has a local as well as a Latin designation, he reasoned, why should a cloud be any different? And since the need to have the clouds' "proper technical names" always at hand connected the earlier complaints made about Howard's chosen language with the complaint made about dictionaries in the *Gentleman's Magazine*, the joint solution to the problems appeared obvious. Forster approached the editors of the *Encyclopædia Britannica* and won their approval for the new translations, so much so that they were persuaded to include them in the forthcoming supplement to the sixth edition, which was due to be published in the early 1820s.[33] This would ensure that the translations were granted the kind of profile they deserved.

But when Howard heard that translations of his terms were due to appear in the most famous encyclopedia in the world, he was appalled all over again. Something needed to be done to limit their likely impact, and his response took the form of an appeal to his readers in the introduction to *The Climate of London*, in which he strongly urged them to resist the suggested terms:

> I mention these in order to have the opportunity of saying that I do not adopt them. The names for the clouds which I deduced from the

> Latin are but seven in number, and very easy to remember: they were intended as *arbitrary terms* for the *structure* of clouds, and the meaning of each was carefully fixed by a definition: the observer having once made himself master of this, was able to apply the term with correctness, after a little experience, to the subject under all its varieties of form, colour, or position. The new names, if meant for another set of arbitrary terms, are superfluous: if intended to convey in themselves an explanation in English, they fail in this, by applying only to some part or circumstance of the definition; the *whole* of which must be kept in view to study the subject with success. To take for example the first of the modifications—the term *Cirrus* very readily takes an abstract meaning, equally applicable to the rectilinear as to the flexuous forms of the subject. But the name of *Curl-cloud* will not, without some violence to it's *obvious* sense, acquire this more extensive one; and will, therefore, be apt to mislead the learner, rather than forward his progress.

These were the words of a scientific worker who had already seen the need to protect the future of his own contribution. He knew that a rival scheme, even one in the form of a translation, could do serious damage to the life of his achievement, and he felt that the attempt was not what he might have expected from a friend. He had already survived the near miss of Lamarck's failed attempt, as he had successfully seen off Bostock's, and he was determined to survive this one, too. But there was a more urgent reason to resist the appeal that a translation of his terms might have made, beyond the sense of threat to his personal investment: the adoption of a local language of clouds would militate against the wider understanding of such obviously global phenomena:

> But the principal objection to English, or any other local terms, remains to be stated. They take away from the nomenclature its present advantage of constituting, as far as it goes, an universal language, by means of which the intelligent of every country may convey to each other their ideas without the necessity of translation. And the more this facility of communication can be increased, by our adopting by consent uniform modes, terms, and measures for our observations, the sooner we shall arrive at a knowledge of the phenomena of the atmosphere in all parts of the globe, and carry the science to some degree of perfection.[34]

The future of world meteorology, in other words, would depend on a common store of language, and as that language was unlikely to be English, Howard felt that these strangely worded translations were as unhelpful as they were unwelcome, despite their being the work of a valued friend.

But his friend's response to this public rebuttal was to go on the defensive, repeating his translations (in direct opposition to Luke Howard's wishes) in the third edition of his *Researches about Atmospheric Phænomena*, as well as in a strangely heterogeneous almanac that he published in 1824, soon after his conversion to Catholicism.[35] This later work, apart from containing a good deal of information on the feasts and festivals of the saints, was notable for deploying the English cloud terms without any reference to the Latin. So Forster was now openly promoting a rival typology of clouds, and typical passages, such as "the change from Curlcloud to Wanecloud, and indeed the great prevalence of the latter cloud at any time, must be regarded as an indication of an impending fall . . . we have seen little Stackenclouds form and disappear in the space of a few minutes; while the Curlclouds form, change their figures to spots of Sondercloud, and disappear," only confirmed Howard's fear that he had stung his young friend into rebellion.[36]

By this time the long-awaited supplement to the *Encyclopædia Britannica* had appeared, and from there, albeit slowly and patchily, Thomas Forster's translations began to spread. Howard was put in an awkward position, but fortunately he was disinclined to make any further effort to resist them. Had he done so, they might well have become far better known than in fact they did. When others picked up on them, it was always with a sense of caution: Admiral Smyth, for example, included them in his *Sailor's Word-Book*, although he seemed more inclined to give precedence to the original Latin.[37] Thomas Milner, too, diplomatically gave both in his *Gallery of Nature*, noting that "Mr Luke Howard's ingenious scheme is now universally adopted, which will be briefly given, placing Mr Forster's English names beside the Latin nomenclature of the former."[38] Others, however, grew increasingly outspoken in their defense of Howard's set of terms. Henry Stephens of the Royal Society of Edinburgh, for example, after considering Forster's contribution to the subject, rejected it outright, concluding that "we must take the nomenclature which the original and ingenious contriver of the classification of clouds, Mr Luke Howard of London, has given."[39]

Howard's was the only clear means by which to name the clouds, and this eventually became the majority view, confirmed when Forster's strange translations were dropped from the subsequent edition of the *Encyclopædia Britannica.* They had failed to answer the need for which they had been made. Instead of the clarification it intended, the rival terminology had only sown further confusion among its users. If it was hard enough to remember one set of terms, it was well-nigh impossible to deal with two; and the Latin set, if strange at first to some ears, nevertheless possessed the clear mark of authority.

Reluctantly, Forster conceded defeat in the end, and the whole episode ended up as little more than another curious distraction in the history of classification. He and Howard soon went on to settle their differences during the first few meetings of the newly founded Meteorological Society of London.[40]

Howard's original essay, meanwhile, had begun to appear in translation in other European scientific journals. The great Swiss scientific activist Marc-Auguste Pictet prepared a translation for his Geneva-based periodical the *Bibliothèque Britannique*, which he had founded in 1796 as a means of keeping continental readers up to date with the latest research from Britain. He and his journal were to rise in importance during the Napoleonic Wars, as they kept fragile lines of communication open for the benefit of international science. Pictet had sought out Luke Howard during a visit to London in 1802, just before the lecture on clouds was delivered, and he had been pleasantly entertained by his host in the observatory built on the top floor of the house at Plaistow. From there, according to Pictet, "every atmospheric modification was visible," and he was much impressed not only by the view but by Howard's comfortable combination of domestic and scientific life. When the translation of the essay appeared in 1804, it was offered up as a testimony to the "*persévérance et sagacité*" of the cloud-haunted young English scientist.[41]

The following year the Geneva translation was translated into German by Ludwig Wilhelm Gilbert, the professor of physics at the University of Halle, for publication in his monthly scientific journal, the *Annalen der Physik*.[42] As did Pictet's publication, Gilbert's journal solicited the majority of its articles from British scientists, a distinctive group regarded by their contemporaries as the most innovative researchers in the world, especially since events on the European mainland, caused by the Revolution in France and the subsequent rise of Napoleon, had entirely disrupted the patterns of intellectual life on the Continent.

Ten years later, Gilbert published a special meteorological issue of the *Annalen der Physik* that featured as its centerpiece a fuller version of Howard's essay. This version had been translated from the 1811 article from Nicholson's *Journal*, "The Natural History of Clouds," and was thus at a third remove from the 1803 original. The meteorology issue of the *Annalen* was concerned with assessing the impact of Luke Howard's wider contribution to cloud physics, and in particular his theories of electricity, as well as the descriptive benefits bestowed by the terms of the classification itself. A series of articles, one of which was written by Thomas Forster, the translator, testified to the lasting value of Luke Howard's meteorology.

By this time Howard was regarded as the science's greatest living exponent—authoritative, widely published, and cited wherever the subject was discussed. What's more, a groundbreaking achievement was forever to be associated with his name. Yet alongside the now familiar encomiums and praise for the Englishman, which, as always, caused him acute embarrassment and disquiet, the issue of translation arose once more.

By the late eighteenth century most European countries had abandoned the use of Latin for scientific treatises, and translation into vernacular tongues had become an important element in the transmission of new ideas across national and linguistic boundaries. Latin remained the global language of natural classification itself, but actual scientific writing was now composed almost exclusively in modern local languages. As Gilbert pointed out, since the majority of German readers had little or no Latin vocabulary, even a few isolated terms, such as "cirrus" and "stratus," might well prove too much for their memories to bear. In spite of what we might imagine about the European past, only a tiny, educated minority was or had ever been remotely familiar with scientific Latin or Greek. Then, as now, dead languages were not widely read or understood, even among the generally well educated. And so, just as Thomas Forster was about to do in England, Gilbert made an attempt to find modern equivalents for Howard's terms. Cirrus, for example, was translated as "Die *Locken-* oder *Feder-Wolke*" ("hair- or feather-cloud"); cumulus as "Die *Haufen-Wolken*" ("heap-cloud"); stratus as "Die *Nebelschicht*" ("fog-layer"); and cirrocumulus as "Die Schafwölkchen" (an already existing German term meaning "sheep cloudlets").[43] But, unlike Forster, Gilbert expressed profound reservations about the value of demotic translation, and he soon gave up the attempt to provide appropriate German terms. Cirrostratus in particular had proved stubbornly resistant to his efforts, and it was this that had convinced him of the futility of his translations.

His famous compatriot Johann Wolfgang von Goethe was in complete agreement on the matter, writing that the names "should be accepted in all languages; they should not be translated, because in that way the first intention of the inventor and founder of them is destroyed." Goethe went on to make a characteristically vigorous defense of Luke Howard's Latin terminology:

> If I hear the term *stratus* I know that the conversation is about the scientific modification of clouds, and we talk of this with those only who know something of the subject. In the same way, by retaining such a terminology the intercourse with foreign nations is rendered more easy. Let it also be considered that by this patriotic purism of style nothing is gained; for since it is known, as it is, that clouds only are being discussed, it does not sound well to speak of 'Heap Clouds,' &c. and to keep repeating the general when speaking of the particular. In other scientific descriptions this is expressly forbidden.[44]

Goethe had spoken, the point was taken, and the nomenclature remained in Latin throughout Europe, much to the irritation of Forster and Bostock, although more particularly to the irritation of Jean-Baptiste Lamarck, whose own attempts at classification had been so unceremoniously truncated. Now he could only watch from the sidelines as Luke Howard's words flowed unimpeded into the hearts and minds of his fellow Europeans.

Beyond Europe, meanwhile, meteorologists were proving a little slower in applying the new terms. Weather correspondents in North America, for example, continued throughout the 1820s to rely on longhand descriptions such as "light broken clouds" or "streaked masses" in their journals.[45] Early American meteorology was understandably more concerned with measuring the practical effects of climate on fisheries and agriculture than in arguing over the niceties of expression. But once Howard's terms were introduced to the country in 1830 through the first *Encyclopædia Americana*, there was (as had already been the case in Western Europe) no stopping them in their tracks.

The contents of this first American encyclopedia were based largely on direct translations of a German-language dictionary that had recently been published to great acclaim in Europe. Most of the translations were accurate enough, but in the course of the lengthy entry for "cloud," a revealing misnomer can be found: "the natural history of clouds, not as respects their chemical structure, but their forms, their application to meteorology, and a knowledge of the weather, has been well treated by Lucas Howard, in his Essay on Clouds."[46] A summary of the Latin classification then followed. Howard's name had remained in German, but the cloud terms themselves, still in Latin within the German text, were then adopted en bloc for American usage without any need for retranslation. Howard's instinct in suggesting from the start that a Latin-based nomenclature would be able to survive any journey through a multiple-translation exercise was thus proved entirely correct, the retranslation of his Christian name to "Lucas" being an instructive case in point. It is intriguing nonetheless to imagine how the American translator might have chosen to render "Die *Locken*- oder *Feder-Wolke*" into English.

The names for the clouds soon caught on among American meteorological correspondents. General Martin Field of Fayetteville, for example, had been contributing for some years to the *American Journal of Science and Arts* before he first employed the Howard terms in 1833. His engagement with the classification, it must be said, spoke more of relish than of accuracy. In an early entry he described how, as he was out walking at around ten o'clock on a sunny August morning in Vermont, "bright *cumulous* clouds of a very slender form, arose from north west to south west. When these clouds had risen to the height of 20° above the horizon, nearly at the same time, *strata* clouds were formed, which lay horizontally upon, and capped the cumulous, and they immediately assumed the form of Mushrooms."[47] Field's misspellings suggest that he may have heard the new names and descriptions spoken aloud, rather than seeing them himself in print. Perhaps, indeed, he had heard them at a lecture, or had had them pointed out to him by a fellow meteorologist in the field. Despite the plethora of printed sources, word of mouth was still the main agent of intellectual exchange, particularly in regions such as the early United States of America.

Other, more idiosyncratic forms of American weather classification were still being devised, meanwhile, and (perhaps not surprisingly) the Turkish-born botanist Constantine Samuel Rafinesque was responsible for a major proportion. Rafinesque had first come to America in 1802, at age 19, for the purposes of travel and research, and after sojourning back in Europe for a time he had returned to make the New World his permanent home. Although his interests were wide-ranging, his main scientific preoccupation was botanical classification, and over his short but event-filled lifetime he attempted to contribute several thousand names and amendments to the lexicon. He was by all accounts a brilliant teacher, and he was appointed professor of natural history at Transylvania University in Lexington through the influence of a well-positioned friend. From there he traveled far and wide in pursuit of his mercurial researches.

But over time he developed a reputation as something of a loose cannon. "He never grew accustomed to the behavior and ideas of ordinary men," as the *Dictionary of American Biography* concluded, and he "never acquired the orderly methods and mental attitude of the trained scientist. Much of his personal suffering and of the ineffectiveness of his work can be traced to this unconquerable innocence." Certainly, by the time his last and strangest work, *The Amenities of Nature,* was posthumously published in 1840, there were few who were prepared to take him seriously. Looking at the book, it is not hard to see why. In it, Rafinesque attempts a thoroughgoing reclassification of natural knowledge in its entirety. The time had come to institute order among the teeming ranks of science, he claimed; all classifications needed themselves to be classified. The task, moreover, had fallen to him alone to complete. His resolve was as impressive as his ego and, among the dozens of new classes and new divisions of science that he proposed, the following are of particular interest:

> ATMOLOGY, science of the Atmosphere.
>
> 1. *Aerology*, science of the air— *Aerognosy*, the physics of it— *Aerography*, description &c.
> 2. *Meteorology*, science of meteors— *Anemology* of winds— *Nephology* of clouds— *Yetology* of rains— *Phosology* of luminous meteors— *Sterology* of solid meteors &c.[48]

Another of his projects, according to his obituarist, was a classification, in full "natural history style," of "*twelve new species of thunder and lightning!*"—each named, arranged, and described according to form.[49]

Exuberant though they were, Rafinesque's redefinitions, like Lamarck's *Annu-*

aires of the early 1800s, or Forster's and Bostock's translations of the 1810s, met with nothing but bemusement from his peers. Rafinesque, whose sanity was in serious doubt toward the end of his life, was fated to become yet another fallen foot soldier of atmospheric science. The nineteenth century was, after all, a century that was to favor global standardization over idiolectic caprice, and, as Samuel Johnson once sadly commented on the baffling *Tristram Shandy*, nothing so odd will do for long.

But by 1840, the year of Rafinesque's death, full and accurate cloud tables were being kept by an increasing number of natural philosophers, such as the great storm chaser Elias Loomis of Western Reserve College in Ohio. Loomis also joined forces with his European colleagues in rejecting the use of "nimbus" as a cloud category in its own right, as can be seen in the table opposite.[50]

By 1857 the Smithsonian Institution, established in Washington some eleven years earlier, had decided to produce an official "Register of Meteorological Observation." This took the form of a large sheet of paper divided into columns, with instructions printed on the back on how to fill them in. The cloud column was divided into two parts: "Course of Higher Clouds" and "Kind of Clouds," the latter to be entered with the appropriate abbreviations, as given: "*St.* Stratus: *Cu.* Cumulus: *Cir.* Cirrus: *Nim.* Nimbus: *Cir. St.* Cirro-Stratus: *Cu. St.* Cumulo-Stratus: *Cir. Cu.* Cirro-Cumulus."[51]

So by the time the United States Signal Service established the Weather Bureau in 1870, American meteorological discourse had long been united with the usages of the wider world.

MONTHS.	9 A.M.						3 P.M.					
	CIRRUS.	CUMULUS.	STRATUS.	CIRRO-CUMULUS.	CIRRO-STRATUS.	CUMULO-STRATUS.	CIRRUS.	CUMULUS.	STRATUS.	CIRRO-CUMULUS.	CIRRO-STRATUS.	CUMULO-STRATUS.
March,	10	3	26	5	7	11	8	11	21	8	5	11
April,	6	12	32	2	8	6	5	14	25	6	6	10
May,	5	7	17	3	8	19	4	14	11	5	6	23
June,	11	15	21	7	9	10	10	32	15	7	6	10
July,	13	28	16	13	3	4	6	47	11	10	4	8
August,	12	31	15	11	5	8	7	51	11	8	3	7
September,	7	16	14	5	8	14	3	21	7	7	5	25
October,	4	13	28	6	11	18	7	18	29	2	7	12
November,	4	7	36	3	7	20	5	10	31	7	10	18
December,		3	66	1	5	9	2	3	56	1	9	12
January,	3	2	57	2	10	13	3	3	51	5	10	16
February,	9	2	37	5	12	13	7	2	33	7	12	12

American clouds in 1840

Chapter Nine

Fame

Fame is, alas! a tinsel shred
Bound on the temples of the dead,
Full dearly bought with peace of mind
To envy and to care resign'd.

Luke Howard, 1808[1]

Many versions of Luke Howard's essay on clouds were printed and reprinted during the years immediately after its first appearance, but not all of them would necessarily have been authorized or paid for. Literary piracy was rife throughout the eighteenth and nineteenth centuries, particularly when a well-known piece of writing was involved, and authors and their publishers found themselves continually embroiled in royalty-chasing lawsuits. Alexander Pope had once infamously administered a dose of near fatal poison to the pirate-publisher Edmund Curll, who had dared to print a purloined edition of some of Pope's satirical court poems. Their feud burned on for another three decades, fanned by the pages of an incendiary pamphlet war. And for every author who, like Pope, sought physical recompense for their injuries or their loss of earnings, there were dozens of others who would have loved to emulate them, although Edmund Curll never repeated the mistake of accepting a drink from the hand of a poet.

But however badly it may have hurt individual authors, piracy was merely one of the many historical vehicles for successful intellectual dissemination. Whether by sanctioned means or not, new ideas and new expressions have always managed to spread themselves with extraordinary rapidity around every corner of the listening world. But as long as piracy remains, whether as an affliction or a temptation, the right names are not always guaranteed to attach themselves to the right ideas. Even the gentle John Claridge, self-styled spokesman for the Shepherd of Banbury, was capable of practicing a series of minor deceptions for gain.

Little is known of John Claridge's life, and what is known is almost certainly apocryphal. He claimed, for instance, that forty years of firsthand observations made "in the open air, and under the wide spread Canopy of Heaven" supplied the material for his best-selling *Shepherd of Banbury's Rules to judge of the Changes of the Weather*, a book first published in 1744 and many times thereafter.[2] The Shepherd, he insisted, employed "the sun, the moon, the stars, the clouds, the winds, the mists, the trees, the flowers, the herbs, and almost every animal with which he was acquainted" as "instruments of real knowledge" and the book, blending science with rural sentiment, proved an enduring favorite over the following two centuries.[3] The fictitious "shepherd" entered the popular consciousness as a symbol of simple rural wisdom and prescience. Even Luke Howard conceded that the modern naturalist is "still obliged to yield the palm in the science of prognostics to the shepherd, the ploughman, or the mariner who, without troubling his head about things, has learned, by traditional experience, to connect certain appearances of the sky with certain approaching changes."[4] Their knowledge of the weather, including the knowledge of Claridge's own shepherd, was revered for its rustic authenticity.

But a letter to the *Gentleman's Magazine*, printed in the May 1748 issue, pointed out that the Shepherd of Banbury, whoever he might be, was not exactly what he seemed. Claridge, it appeared, had been guilty of appropriating most of his "lore" not from the wisdom of a watchful shepherd but from the pages of an earlier book:

> Mr Urban,
>
> *A Rational Account of the Weather, by the Rev. Mr* Pointer, was published in 1738; and *The Shepherd of Banbury's Rules*, &c *by* John Claridge, in 1744.
>
> The Observations of the Shepherd, contained in the latter Treatise, are said to be grounded on no less than 40 years experience; but by comparing them with those contained in the former account, I find almost all his observations to be transcrib'd *verbatim* from it.[5]

The claims of the characteristically pseudonymous correspondent (who styled him- or herself as "Stalbrigiensis") were followed up by twenty-six examples of plagiarisms from Pointer, the sole aim of which was clearly to expose John Claridge's borrowings to the widest possible audience:

> Clouds small and round, like a dappley-grey, with a north wind,—fair weather for 2 or 3 days. *Shepherd.*
> *Clouds appearing white like fleeces of wool, scattered about in the sky, are another sign of fair weather.* Pointer.
>
> Large like rocks,—great showers. *Shepherd.*
> *Clouds appearing like rocks or towers, signify great showers.* Pointer.
>
> If small clouds increase, much rain. *Shepherd.*
> *If small clouds grow bigger and bigger in an hour or two, they signify a great deal of rain.* Pointer.
>
> If large clouds decrease,—fair weather. *Shepherd.*
> *If great clouds separate, waste off, and grow smaller and smaller, this signifies fair weather.* Pointer.[6]

And so on for another two incriminating pages. John Pointer's earlier (and wordier) book had been plundered like a quarry for material to be offered up as hard-won weather lore. The shepherd was no more than a fictional device to mask the unacknowledged borrowings from Pointer.

But this exposure did little to discourage Claridge's sales or the frequent reprintings of the book over the rest of the century. In fact, the episode only went to show how buoyant the midcentury market already was for books about weather and clouds. And by the end of the century that market had grown out of all proportion amid a burgeoning sea of print.

A reader walking through the streets of London had a wide choice of venues at which to pause to browse or buy a copy of any of the publications in which Luke Howard's words appeared. At Cadell and Davies on the Strand, perhaps, or at Longman & Rees on Paternoster Row; or there was Vernor & Hood's on Poultry, or Harding's more aloof establishment at 36 St. James's Street. The title page of the *Philosophical Magazine* listed nine approved retailers in central London, plus a further two in Scotland and another in Dublin. The name of a bookseller in Hamburg ("W. Remnant") was appended for the benefit of northern European readers. And any of the other books and periodicals that carried extracts from Howard's essay might also be found at these or any of the dozens of other shops and stationers scattered throughout the print-mad city: at Debretts or Hatchards on Piccadilly, Bells on Oxford Street, Richardson's of Cornhill, or, best and biggest of all, James Lackington's imposing "Temple of the Muses" in Finsbury Square. Lackington's famous bookshop, with its great domed classical interior, was by all accounts a book collector's dream. It swam with printed matter: bound and unbound books, essays, periodicals, song sheets, and stationery drifted in a restless sea of words, to which Howard had now added his small but influential share. Anyone might read his essay on clouds now that it was out in the world, just as anyone might hear his words now that they were spoken aloud into the listening air.

The drawback of all this, however, was fame. Howard had soon found himself feted as a scientific celebrity, courted with worldly flattery and acclaim, and he greeted this unexpected change in his fortunes with something approaching trepidation. He valued his success, of course, and was pleased to have made such a lasting contribution to science, but the personal recognition that came alongside it made him very uncomfortable indeed. Such worldliness did not sit well with his strongly held Quaker convictions, and in 1808 he composed a subdued, semimystical poem entitled *My Ledger*, in which he meditated on the qualities of fame in the lines reproduced at the head of this chapter: "*Fame* is, alas! a tinsel shred/Bound on the temples of the dead . . ."

It was not unusual for scientific writers to express their thoughts in verse: Erasmus Darwin, Gilbert White, and Humphry Davy had all routinely done so in earlier decades, and others, such as Thomas Lovell Beddoes and James Clerk Maxwell, would go on to do the same in later years. Some of their poetry was magnificent, although most of it was less so. Davy, for example, had written with characteristic eloquence of scientists as "the sons of nature," exploring "the tranquil reign of mild Philosophy" until rapture filled their souls, while Maxwell, equally characteristically, had mounted bad-tempered verse objections to recent theories of molecular evolution. "Hail Nonsense!" he wrote:

> From thee the wise their wisdom learn,
> From thee they cull those truths of science,
> Which into thee again they turn.
> What combinations of ideas,
> Nonsense alone can wisely form!
> What sage has half the power that she has,
> To take the towers of Truth by storm?[7]

For Howard, though, the move to confessional poetry marked the end of a period of painful adjustment, during which his newfound fame, the end of his working partnership with William Allen, and the death of a child in infancy had all occurred in close proximity. The loss of their infant daughter Mariabella, "a lovely little child" who was named after her mother, hit the young family particularly hard. She died of whooping cough at eighteen months of age, and the funeral at Barking was the worst day of their lives so far, bringing back memories of Robert Howard's grief at the loss of his sons in the early 1790s.

It was a time of great emotional upheaval for the family, and while Bella went to stay with some cousins for a rest, Luke sought a temporary refuge from his various accumulating stresses. In the summer of 1807 he and William Allen took a walking holiday together in the Lake District of northern England, partly to discuss their future plans, partly for well-earned rest and recuperation. Allen, too, was exhausted. He had just been elected a Fellow of the Royal Society, having found his own fame as a scientific lecturer, with sellout shows at the Royal Institution and a regular public series held at Guy's Hospital in Southwark. Like Howard, he tried to live at a remove from his newly won acclaim, reminding himself in late-night diary entries to "be on my guard against the world's flattery and applause," while praying to be "preserved from the inordinate love of science."[8] The temptations of fame and a rising reputation proved sorely trying to their modest Quaker spirits, especially for the widowed William Allen, who faced an appreciative public audience every evening. Following his earlier experiences with hallucinogenic drugs, Allen's temptations seem once more to have been given in to, since he was later lampooned in the press for the frequency of his serial marriages. But his crises of confidence were all too real, brought about in part by his success. They were scrupulously recorded in his personal diaries, in a painful chronicle of an escalating stage fright:

> 3rd December 1801: At seven o'clock I gave my first lecture on chemistry. I got through beyond my expectations, but I was very low about it before I began.
>
> 13th February 1802: Rose early — getting ready for experiments at the Hospital — I felt distressingly low and anxious — it began and ended with loud plaudits.
>
> 21st October 1802: First lecture at the Hospital this season . . . I am very anxious and fearful.[9]

Once the two friends had decided that they needed some time away from London, the lure of the Lakes in summertime was not to be resisted. They headed north with a lightening step and a sense of adventures to come. They could discuss the separation of their business interests in the congenial setting of hills and lakes.

The split had its origins in 1805, when Howard moved the manufacturing laboratory from Plaistow to larger premises in nearby Stratford, although he and his family continued to live in Plaistow. It was clear not only that the laboratory's value had massively outgrown that of the Plough Court pharmacy but that its further expansion was being hindered by the connection. Allen agreed to found another, smaller factory nearer to the headquarters at Lombard Street, while Howard would pursue larger-scale manufacturing as a separate venture under his own name. Thus, on a summertime walking tour in the Lakes, Allen & Howard was amicably ended, and Luke Howard & Company of Stratford was born.

The popularity of Lakeland as a holiday destination had been growing over the previous three decades, a trend given external impetus by the Napoleonic blockade of continental Europe. The Lake poets William Wordsworth and Samuel Taylor Coleridge were among the first to complain about the increasing number of "sallow-faced & yawning tourists" to the area, and in 1805 one had even stumbled to his death near the summit of Helvellyn—the body was discovered three months later with a Claude glass in its hand. Poor Charles Gough (no relation to blind John): the first man to die in search of the picturesque. Wordsworth and Walter Scott both wrote elegies to his memory, and his dog lives on in local legend.[10]

By 1807 tourists to the Lakes were airing their own complaints about the scarcity of beds in summer, but as there was a long-established and thriving Quaker community in the region, Allen and Howard would have had no difficulty in securing lodgings, even at the height of the season. There were certain material advantages to be had in belonging to a spiritual network.

Once ensconced in their lodgings at Keswick, according to their diaries, they did what most nineteenth-century scientific enthusiasts would have set about doing on their holidays: they rose early to go hill-walking, climbing some of the loftiest mountains to enjoy the views; they had picnics on the summits; and they tracked the changes of atmospheric pressure with their portable barometers. Working out their observations, and comparing them with earlier published records, gave the pair "most satisfactory evidence of the accuracy of their calculations."[11]

Luke Howard's Westmorland journal, a small calf-bound notebook now lodged in a cardboard box in the London Metropolitan Archives, is filled with his readings from the barometer and thermometer, calmly noted down amid some of Europe's most spectacular scenery. His account of their barometrical ascent of Helvellyn is one of the journal's highlights. The climb took several hours, but after the pair had taken their twin sets of measurements and enjoyed the unrivaled clarity of the distant views of mountains and lakes, spread before them in glorious sunshine, the face of the weather began to change with a familiar Lakeland rapidity. The fine morning turned to present instead "a fine and truly gratifying view to a meteorologist of large cumuli mixing with the mountains, gliding up the valleys and sailing by us with their round summits beneath our feet while the [superiors?] mixed with scud poured down rain upon our heads."[12] Scud, or stratus fractus, is a smaller, darker "accessory cloud" that hangs below a rain-bearing structure. They had long been referred to as messenger clouds by navigators, farmers, and millers, and the message they brought was rain. Like John Evelyn's interior view of an Alpine cloud, this was one of the sights of Howard's trip, but unlike Evelyn,

Howard had the language—the language that he himself had bestowed—with which to apprehend and describe what he had seen.

After recovering from their drenching with the aid of a generous measure of brandy, the pair began their slippery descent, with Howard suffering several bruising falls "yet without injury to the Barometer which on all the occasions I carried *slung* to my waist . . . holding it before in descending, behind in going up." Even with the recent memory of Charles Gough's death on the same well-worn section of the mountainside, Howard's only anxiety was over the "danger to my Barometer." Nothing else, it seems, could diminish his appetite for meteorological adventure, and the ascent of Skiddaw a few days later elicited Howard's same exhilaration as he found himself swung high above the dizzying "zone of the Clouds." He had never, he realized, been face-to-face with the distant companions of his youth, but the views on this trip had deepened his horizons and rendered the clouds closer and more majestic than ever.

Scientific pastimes such as these were then at the height of their popularity, with geological, botanical, and climatic attractions acting as a particular spur to the growth of scenic tourism. They had also become, unsurprisingly perhaps, a regular target for the satirists. James Plumptre's 1798 comic opera *The Lakers*, for example, sent a group of scientific amateurs up a hillside, accompanied by "a guide and botanist, 'well acquainted with the *indigenous* plants of the mountains, rocks, and lakes,' " heading for the summits of absurdity, to enjoy their packed lunches amid a view "set off by the empurpling and downy hue of the clouds."[13] The whole thing—dedicated sarcastically "to tourists in general"—could as easily have been a description of the antics of Allen and Howard, or even those of Wordsworth and Coleridge, whose own obsession with the climate of the Lakes saw them piling up the hillsides in the pouring rain, pausing only to bless the bounty of "his Supreme Majesty's Servants, Clouds, Waters, Sun, Moon, Stars, etc.," all of whose anacoluthic changeability appealed famously to their romantic sensibilities.[14]

The Lakes were not only a scene of popular scientific tourism. They were also strongly associated with a group of English writers, the Lake poets, who were distinguished by their interest in new ideas and new classifications, especially classifications of affect. For them, the new names for the clouds took their place alongside other recent critical terms, such as the uppercase "Sublime" and "Picturesque," both of which were also newly codified categories of visual experience. All such terms were put to work on the slopes overlooking the Lakes, whether by meteorological enthusiasts such as Allen and Howard, or by poetic visual thinkers such as Wordsworth and Coleridge, who pursued languages of taxonomy with great seriousness. Like any natural scientist of the age (and Coleridge, the friend of Davy, certainly regarded himself as scientific in outlook), they felt the need to redress uncertainties of meaning wherever they might occur.

It was not always an easy responsibility. Dorothy Wordsworth recorded a painful encounter between Coleridge and a fellow scenic tourist at the falls of Cora Linn. The stranger set it off by observing "that it was a '*majestic* waterfall' ":

> Coleridge was delighted with the accuracy of the epithet, particularly as he had been settling in his own mind the precise meanings of the words grand, majestic, sublime, etc., and had discussed the subject with William at some length the day before. 'Yes, sir', says Coleridge, 'it is a *majestic* waterfall'. 'Sublime and beautiful', replied his friend. Poor C. could make no answer.[15]

For poets such as Wordsworth and Coleridge ("Poor C."), visual categories were important means with which to understand the operations of the world. To witness the misuse of those categories was a painful business because it went to the very heart of their intellectual endeavor. The processes of the intellect needed to be sympathetically matched to the intricate processes of nature, and language was the most important means of achieving such a match. Coleridge once described the Lakeland panorama as a scene in which "mists, & Clouds, & Sunshine make endless combinations, as if heaven & Earth were forever talking to each other," and this idea, that we might somehow listen in on the great conversations of nature, had been as important for Howard as it had been for Coleridge.[16] It was a major factor in the success of his classification of clouds. Language, therefore, for these Romantics, was not something to be lightly misused. Coleridge's friend, like Forster's translations, conveyed little more than a clumsy tribute to the power of the original.

Howard, who made watercolor sketches of the summer clouds as they formed over the mountains of the Lakes, must have been made increasingly aware of the enabling qualities of the language he had given to the world. It had the unambiguity sought by thinkers such as Coleridge, yet it breathed with the liberating promise of mutability.

This promise of mutability was the classification's rallying cry, for some of Howard's romantic contemporaries were growing impatient with what they saw as the life-denying conventions of the European scientific Enlightenment. A noisy reaction, albeit a marginal one, was building against the materialist emphasis laid by science on the unraveling of the processes of nature. The horrors of *Frankenstein* were just around the corner and would shortly give both a name and an iconography to this growing antiscientific sentiment. "My reign is not yet over," as the nameless monster taunts his unnatural maker. "Come on, my enemy; we have yet to wrestle for our lives."[17]

This fearful sentiment was nowhere better expressed than at a memorable evening in December 1817, when William Wordsworth, John Keats, Charles Lamb,

Watercolor of clouds by Luke Howard, c. 1807

and Thomas Monkhouse (Wordsworth's cousin-in-law and unofficial agent) had dinner together at 22 Lisson Grove North, the London home of the history painter Benjamin Robert Haydon. Haydon at the time was preoccupied with what was to become his best-known painting, *The Entry of Christ into Jerusalem*, which featured, among the crowd of spectators at the central scene, the portraits of three modern "geniuses": Wordsworth, Newton, and Voltaire. As Haydon himself commented, the three illustrious heads made "a wonderful contrast," and he was soon to add more to this painted pantheon.[18] His guests, however (assembled for dinner in the painting room), were more disturbed than pleased by the combination. According to Haydon's diary, as the evening wore on, an argument over the imaginative claims of science and poetry, as personified by the trio of portraits, developed among the five friends. The finale was encouraged by the enlivened Charles Lamb:

> Lamb soon gets tipsey, and tipsey he got very shortly, to our infinite amusement. He then attacked me for putting in Newton, 'a Fellow who believed nothing unless it was as clear as the three sides of a triangle.' And then he and Keats agreed he had destroyed all the Poetry of the rainbow, by reducing it to a prism. It was impossible to resist them, and we drank 'Newton's health, and confusion to mathematics!' . . . it was an evening worthy of the Elizabethan age.[19]

Lamb's impassioned isolation and rejection of the scientific view of nature was an unusual position for an early-nineteenth-century intellectual to take, even in his cups, and its force made a particular impression on the 22-year-old Keats.

Keats had already turned his back on a medical career that had occupied half his adult life, and had just published his first volume of poems. Despite its cool critical reception, its author had begun to formulate in his mind the hypnotic idea of poetry as a lifelong vocation. Such a life would involve the rejection of anything that he considered nonpoetic and non-life-affirming, or which threatened the value of poetry's special insight. This was a long way from the belief in scientific enlightenment that had marked the preceding generation. But truth, for Keats, was a glimpse of unclouded light, not the broken ray of Newton's prism. Truth derived from "being in uncertainties, Mysteries, doubts, without any irritable reaching after fact & reason."[20] How could this new outlook, far removed from what it saw as the pedantry and shortsightedness of natural philosophy, be reconciled with anything like a scientific outlook?

With thoughts such as these still fresh in his mind, the young poet was ready both intellectually and emotionally to occupy the antiscientific position suggested by Lamb during the dispute over the inclusion of Newton. The dinner party proved to be a potent, resolving moment for Keats, and in the days immediately following the altercation (with its heroic toast against everything empirical), he began to develop his defense of the rainbow in what were to become the best-known lines from "Lamia":

> Do not all charms fly
> At the mere touch of cold philosophy?
> There was an awful rainbow once in heaven:
> We know her woof, her texture; she is given
> In the dull catalogue of common things.
> Philosophy will clip an Angel's wings,
> Conquer all mysteries by rule and line,
> Empty the haunted air, and gnomèd mine—
> Unweave a rainbow, as it erewhile made
> The tender-personed Lamia melt into a shade.[21]

The refusal to recognize any spiritual worth in the Newtonian approach to nature, condemned by the poet as "cold philosophy," is now familiar, but at the time it was not a widely heard sentiment. In fact, the earlier generations of poets had seen nature transfigured and unveiled by the light of Newton's prism. Richard Blackmore's "Creation" of 1712, Mark Akenside's "Hymn to Science" of 1739, or the best known of all, James Thomson's "Poem Sacred to the Memory of Sir Isaac Newton" of 1727, expressed reverence for the insights granted by the new science:

Even *Light itself*, which every thing displays,
Shone undiscover'd, till his brighter Mind
Untwisted all the shining Robe of Day;
And, from the whitening undistinguish'd Blaze,
Collecting every Ray into his Kind,
To the charm'd Eye educ'd the gorgeous Train
Of *Parent-Colours*. First the flaming Red
Sprung vivid forth; the tawny Orange next;
And next delicious Yellow; by whose Side
Fell the kind Beams of all-refreshing Green.
The pure Blue, that swells autumnal Skies,
Ethereal played; and then, of sadder Hue,
Emerged the deepened Indigo, as when
The heavy-Skirted Evening droops with frost;
While the last Gleamings of refracted Light
Died in the fainting Violet away.[22]

Here was science and poetry in colloquy rather than in opposition, supporting each other's imaginative appeal; but it was this very colloquy that was to be irreparably shattered by the attitude of Keats and his circle.

The novel idea that poetry might enjoy a higher cultural worth than science was intended to be an expression of the limits of empiricism. Creative insight (itself a romantic idea) began to be promoted in opposition to the efficient mensuration of the scientists. Poetry was a vocation to be embraced by "literary" souls, while science was a priesthood to be shunned. However much this was a travesty of real intellectual history, the later generation of Romantics came to view the empirical preoccupations of the natural philosophers as at best unsophisticated, and at worst deadeningly boorish. Keat's image of the rainbow, divided by the cold head and closed heart of science, was offered up as a tragic emblem, torn and fluttering in the violated sky. Here, then, was a negative standard, an image of the manifold failure on the part of the scientific mind either to behold beauty or to be moved by it. And beauty, for Keats, was truth.

But Howard and his clouds, ineffable as they seemed, would remain untouched by such artistic or ethical censure. Instead he would be credited with the creation of a system that offered necessary precision to its users while allowing for a margin of magnificent change. He reaffirmed meteorology as a science of contemplation, which, for Keats, would hardly have included it in the sciences at all. He had given the world a life-affirming language that celebrated the poetry of nature. And in bestowing this gift, Howard had helped redefine the romantic spirit of the age.

Portrait of Luke Howard by John Opie, c. 1807

This, however, only brought him more of the fame that he shunned. Following his return from the exhilarating Lakes, he was persuaded to sit for John Opie (1761–1807), a leading society portrait painter, after which he composed his romantic crisis ode in reaction to his newfound celebrity.

The name of Luke Howard had now entered the nacreous realm of fame. Like Humphry Davy, who confessed to hearing "the voice of fame murmuring in my ear," Howard saw his day-to-day life change beyond recognition from the obscurity of the Fleet Street pharmacy.[23] He was now the "accurate" Mr. Howard, "the well-known meteorologist," the "ingenious contriver of the classification of clouds," who had

Defin'd the doubtful, fix'd its limit-line,
And named it fitly—Be the honour thine!

and in so doing he had become one of the best-known players on the British scientific scene. Everybody wanted to meet him, and there is even a distinct possibility that "on the warm evening of 22 July 1813," he made a house call to the novelist Jane Austen:

> As he travelled through Chawton, just before Alton, he would have passed before Austen's dining room window, the outlook of one who was his equal in meticulous observation. Whether they met that day we do not know but it seems possible. Howard was a campaigning celebrity with links to the Lloyd and Barclay families, Quaker bankers. There were Barclays in Alton, and Austen's brother was a banker. After this time, Austen's letters seem full of weather.[24]

The meeting of Austen and Howard is a tantalizing hypothesis for which there remains, sadly, no real documentary evidence. But what is certain is that Luke Howard of Plaistow, the son of a Clerkenwell tinsmith, had, however reluctantly, become someone to be reckoned with.

And the example of his achievement, so luminous and brilliant, was soon to be an inspiration to others working in the field. Admiral Sir Francis Beaufort, for example, the deviser of the scale that bears his name, owed much to the example of the young Luke Howard, although the admiral, as so often was the case in his long and frustrating life, had a far longer fight for acceptance on his hands.

Chapter Ten

The Beaufort Scale

All honour to Beaufort, who used and introduced this succinct method of approximate estimation by scale, expressed in numbers instead of vague words, about the beginning of this century. By the kindness of his family, we have them now before us, in the log of the H.M.S. 'Woolwich' in his own handwriting, dated 1805.

Robert FitzRoy, 1863[1]

As the new generation of poets and painters began to further their claim upon the realms of the imagination, the sciences continued to take ever greater strides out into the material world. And with the expansion of both naval and merchant fleets, with their need for agreed ways of describing what they saw, meteorology was spreading fast from land to sea. There it would find its methods tested anew amid the lash of the stinging salt air, for the successes and failures of the fast-evolving science had begun to preoccupy the minds of the more intrepid travelers as they journeyed toward the farthest shores.

The 21-year-old ship's captain William Scoresby, Jr., for example, voyaged to Greenland on the whaling ship *Resolution* in the summer of 1810, and in the course of his journey he sailed into the history books as the first truly meteorological sea-goer of the modern era. Scoresby, a whaler from his earliest youth, had become an early convert to Luke Howard's nomenclature, having come across it in a copy of the *Philosophical Magazine* in the library at Edinburgh University. It was there, between whaling trips, that he found intellectual refuge in the chemistry and natural history courses conducted by the great Scottish natural philosophers John Playfair and Robert Jameson. The professors were by all accounts delighted to have had a real frostbitten Greenland whaler sitting in their classrooms, and they listened for hours to the younger man's descriptions of the Arctic ice fields, where "the majestic unvaried movements," as he was later to describe them, exercised a powerful impact on his developing scientific mind.[2] The professors, encouraged by Scoresby's enthusiasm and scientific gifts, pressed him to take detailed meteorological, geological, and oceanographic observations for them as assignments to be discussed upon his return. Scoresby was only too happy to oblige.

Scoresby used the full Howard cloud classification in the ship's log of the *Resolution* in his first journey as captain, in 1810, and found it so useful that he published the tables upon his return as a demonstration piece for his friends and peers, for whom it was already on its way to becoming a familiar language.[3] Arctic clouds, as he explained in a later publication, written in response to the grandeur of the scene, "most generally consist of a dense stratum of obscurity, composed of irregular compact patches, covering the whole expanse of the heavens. The cirrus, cirro-cumulus, and cirro-stratus, of Howard's nomenclature, are occasionally distinct; the nimbus is partly formed, but never complete; and the grandeur of the cumulus or thunder-cloud is never seen, unless it be on the land . . . the most common definable cloud seen at sea, is a particular modification, somewhat resembling the cirro-stratus, consisting of large patches of cloud arranged in horizontal strata, and enlightened by the sun on one edge of each stratum."[4] This was the first navigational use of the new nomenclature, and it is only fitting that its champion, like its author, was a young and spirited enthusiast keen to experience the world firsthand. Scoresby, like Howard, had been dominated by his father, and he, too, had sought and found something in the newly prominent science of meteorology to embrace and call his own.

Scoresby's attachment to Arctic exploration—to meteorological, zoological, and magnetic research—was intense and long-lived enough to earn him the friendship and admiration of Sir Joseph Banks as well as fellowships of the Royal Societies of both London and Edinburgh. In 1813 he invented the "marine diver," an apparatus for measuring deep-sea temperature, and with it he was the first to confirm that the depths of the Arctic oceans were warmer than the heights. He was also the first to deduce that plankton was the principal cause of changes in the coloration of the sea.[5] All in all, he seemed fast on his way to a major scientific career.

But in spite of Scoresby's experience, commitment, and obvious talents, he was fated to be disappointed in his hopes. The 1818 expedition to the Arctic, which set out to discover the Northwest trading passage to Asia, was sponsored by John Barrow of the Admiralty, although the idea itself had originally come from a letter written by Scoresby to his supporter Sir Joseph Banks. But when Scoresby made the trip to London to present himself for interview, he instead came face-to-face with the ambition and malice of Barrow, who made it condescendingly clear that a young Greenland whaler such as William Scoresby, Jr., might set his hopes on obtaining an engagement as a pilot at best, but never as commander of the trip; having delivered his judgment, and without waiting for a reply, Barrow, "turning sharply . . . left the room."[6] Incensed by his treatment, Scoresby returned to his ship, from where he wrote to Banks to predict (correctly, as it turned out) that the expedition would never get to the higher latitudes without him. Although Banks appealed to the First Lord of the Admiralty on Scoresby's behalf, his efforts met with obstruction from the ever scheming Barrow, who, while hampering Scoresby's chances of appointment, had stolen his plans for the Arctic expedition and was soon to see them published as his own. So when Scoresby the whaler sailed soon after for Spitsbergen and the Greenland Sea, he sailed alone, angry and disappointed, but with his appetite for exploration undimmed.

Scoresby continued to employ the Howard system on these later lone and self-funded voyages, going out of his way whenever possible to praise the benefits of precision it bestowed. But as he also pointed out, elsewhere in the language of meteorological description things were not going so well. As a navigator, he needed to keep his mind as much on the sails before him as on the cloud structures rising above. But the winds, unlike the clouds, had not yet found their Howard, and Scoresby was not alone in lamenting that he was unable to be as accurate as he needed to be in his weather accounts when he came to write them up in his logbook at the end of each day's adventures:

> In the phenomena of the Winds, however, which I am now about to describe, I cannot be so precise; being able to give a correct idea only of their peculiarities and direction, whilst their relative force, founded on conjecture, I am unable to express otherwise than in the phraseology of the mariner, which, it must be allowed, is somewhat ambiguous.[7]

Ambiguous was the word. The most widespread definitions of the time were to be found delineated in William Falconer's *Universal Dictionary of the Marine*, a book deemed indispensable on its first publication, in 1769. Falconer's list—"if the wind blows gently it is called a breeze; if it blows harder, it is called a gale, or a stiff gale; and if it blows with violence, it is called a storm or hard gale"—offered obvious room for improvement.[8] There were no degrees to Falconer's laconic descriptions, there were no agreements on definitions of terms, yet his scale seemed to be the best on offer and was widely referred to on board seafaring craft. Even Scoresby's own suggested fifteen-point escalation—"Calm, inclinable to calm, light air, gentle breeze, moderate breeze, brisk breeze, fresh breeze, strong breeze, brisk gale, fresh gale, strong gale, hard gale, very hard gale, excessive hard gale, hurricane"—was not much better.[9]

There was more than one reason for worrying about the wind, for there was not just a need for immediate description but also a need to understand historical reports. This lack had already been pointed out by Daniel Defoe in the course of his account of the storm of 1703. Without agreed terms for present use, he argued, how can we make sense of weather descriptions from the past?

> From hence it comes to pass, that such Winds as in those Days wou'd have pass'd for Storms, are called only a *Fresh-gale*, or *Blowing hard*. If it blows enough to fright a South Country Sailor, we laugh at it: and if our Sailors bald Terms were set down in a Table of Degrees, it will explain what we mean.

Stark Calm.	A Top-sail Gale.
Calm Weather.	Blows fresh.
Little Wind.	A hard Gale of Wind.

A fine Breeze.	A Fret of Wind.
A small Gale.	A Storm.
A fresh Gale.	A Tempest.

> Just half these Tarpawlin Articles, I presume, would have pass'd in those Days for a Storm; and that our Sailors call a Top-sail Gale would have drove the Navigators of those Ages into Harbours . . . when our *Hard Gale* blows they would have cried a Tempest; and about the *Fret of Wind*, they would be all at their Prayers.[10]

Here, then, was the historical dimension to the problematic language of wind. As had been the case with the early languages of clouds, the drawbacks of macrodescriptive terminologies were continually exposed, particularly in the reporting of storms at sea. What one century thought of as a vengeful tempest, another might dismiss as no more than a gale. Where was the potential for objective measurement when the simplest phrases lacked agreement? And how might warning signs be recognized early if their wording was subject to variance?

The elements were clearly in need of further narrative control, and Luke Howard's celebrated capture of the clouds by a leap of imaginative language must surely have heralded the imminent capture of those other ancient intangibles, the sail-filling, boat-destroying winds. There was an obvious parallel between the two problems. But Howard had shown how a descriptive solution could be constructed to account for even the most difficult—because the most intangible—phenomena. Following this line of reasoning, Scoresby grew certain that finding a comparable solution for describing the winds could only be a matter of time; and he was right. The young ship's commander Francis Beaufort (1774–1857) devised his famous wind scale in January 1806 while waiting restlessly for his ship, the *Woolwich*, to leave Portsmouth for convoy duties in the mid-Atlantic. He too had no doubt been encouraged by the recent example of Luke Howard's international success.

Like every mariner before him, Francis Beaufort had been struck by the need for an agreed measurement of wind force. The earlier definitions as outlined by William Falconer, or the four-point numerical scales introduced in earlier centuries by the Royal Societies of London and Mannheim, were clearly inadequate. Robert Hooke had included a four-point scale, in his "Method for Making a History of the Weather," for recording "the Quarters of the Wind and its strength" (see page 130). The wind's changes of direction were a constant source of surprise, and part of the problem lay in the period's incomplete understanding of the causes and effects of wind itself. It was known that there was a connection between weather and atmospheric pressure, but the dynamics of this relationship were only just being grasped.

Winds, whether global or local, are produced when pockets of air move around in their attempts to equalize the temperature and pressure differences caused by the uneven spread of sunlight over the earth. These causes had been described in the mid-eighteenth century by the British natural philosopher George Hadley (1686–1768), but what was not yet understood was that the movements of global winds are complicated by the earth's rotation, which serves to deflect the traveling air from its course. This effect, known as the Coriolis effect after the French meteorologist who identified it (Gustave-Gaspard de Coriolis, 1792–1843), steadily increases in influence toward the poles, where, in the words of William Falconer, "a general or *reigning* wind blows the same way, over a large tract of the earth, almost the whole year."[11]

Local convections and breezes, meanwhile, rise up in response to immediate differences in temperature and atmospheric pressure that occur between the sea and the land, or between different levels of air. Because land heats up more quickly than water, the air above it is warmer during the day, causing cooler air to move inland from the seas or lakes to replace it as it rises. At night the situation is reversed, as land also cools more quickly than does the surface of the water, above which warmer air now rises. Similar processes occur in valleys and on hillsides, causing winds to change direction in the evening.

Natural philosophy held reliable description to be the major route toward improved understanding of phenomena, and so, just as in the case of the clouds, an objective, compatible, and descriptive classification of wind force was sought by many. Francis Beaufort was skilled enough—and lucky enough—to end up with the credit for its discovery.

Beaufort first employed his new wind classification in a journal entry dated January 13, 1806. His ship, the *Woolwich*, had once been a sailing man-of-war, but by then it had been converted humblingly to a storeship; Beaufort, who at the age of 31 considered himself a veteran of many years' standing, identified bitterly with the relegation. The complaints in his diary reveal the hurricane force of his violent disillusionment: "To a storeship! Good Heavens! Is it for the command of a storeship that I have spilled my blood, sacrificed the prime of my life, dragged out a tedious economy in foreign climates, wasted my best hours in professional studies . . . For a storeship, for the honour of carrying new anchors abroad and old anchors home! For a ship more lumbered than a Dover packet, more weakly manned than a Yankee carrier." The *Woolwich* was a ship, he concluded, having exhausted himself with invective, "where neither ambition, promotion or riches can be obtained!"[12] He had taken his demotion to its command as a personal humiliation.

But once Beaufort had calmed himself down, he was able to take a more philosophical approach and apply himself, while unoccupied on board his detested ship, to one of his long-standing interests: the science of the weather at sea. This was a subject that in his view deserved "laborious investigation," and he was already keeping meticulous journals of the effects of maritime weather.[13] He recorded cloud cover, temperature, visibility, and rainfall, but it was the phenomenon of the wind that caught his particular attention. He was determined to render it both knowable and quantifiable, and he framed his resolution in a historic journal entry, made in his private notebook rather than in his captain's log. The entry was prefaced with words born of a familiar frustration: "Hereafter I shall estimate the force of the wind according to the following scale, as nothing can convey a more uncertain idea of wind and weather than the old expressions of moderate and cloudy, etc. etc."[14] Ambiguity was the enemy of naval records, but it was soon to find a powerful adversary in Francis Beaufort of

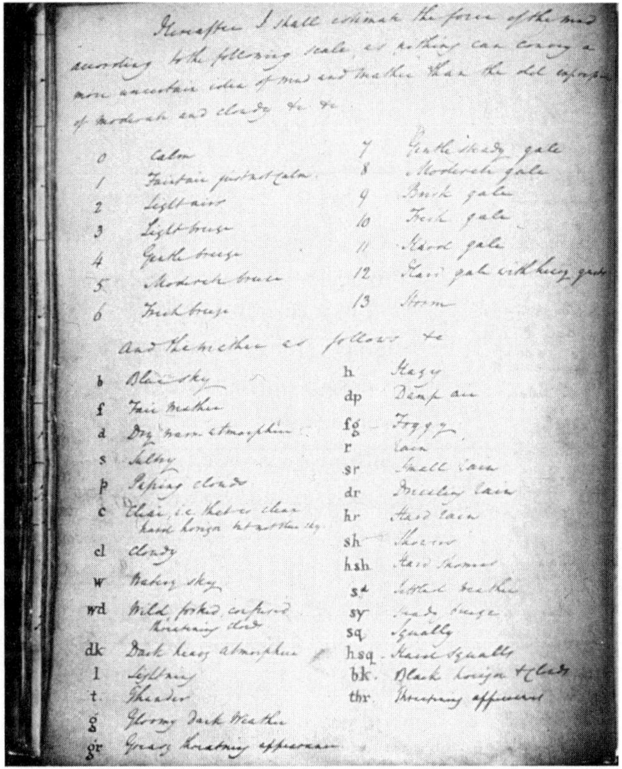

The original Beaufort scale of 1806

the *Woolwich*. This, his first numbered scale, comprised an ascending ladder of fourteen categories of wind force, rising from "Calm" to "Storm" via "Light breeze," "Fresh breeze," and "Gentle steady gale."[15]

A selection of shorthand codifications followed, assigning abbreviations to all weather conditions to be recorded in his log: cl for "cloudy," h for "hazy," thr. for "threatening appearances," and so on.

As it stood, this was not much more than a restatement of already existing wind scales. It was certainly no match for one that had been developed for use on land by a certain Mr. Rouse of Leicestershire, a friend of the celebrated lighthouse engineer John Smeaton (1724–1792). Smeaton was renowned the world over for his design of the third (and greatest) Eddystone Lighthouse, built on the reefs off Plymouth between 1756 and 1759. The elements had wreaked havoc with both the lighthouse's wooden precursors; the first was blown down in the storm of 1703, and the second was destroyed by fire in 1755. They would go on to do the same for Smeaton's own, which was reluctantly dismantled in 1877, having been damaged beyond hope of repair. Less well known than the Eddystone Lighthouse, however, and as yet unknown to Francis Beaufort, was a paper on the powers of water and wind that Smeaton had read to the Royal Society of London in May 1759. The paper, which described the actions of the sails of a windmill, included Rouse's table of wind velocities ranked according to their "common appellations":

Velocity of the Wind.		Perpendicular force on one foot area in pounds averdupois.	Common appellations of the force of winds.
Miles in one Hour.	Feet in one Second.		
1	1,47	,005	*Hardly perceptible.*
2	2,93	,020	*Just perceptible.*
3	4,40	,044	
4	5,87	,079	*Gentle plesant wind.*
5	7,33	,123	
10	14,67	,492	*Pleasant brisk gale.*
15	22,00	1,107	
20	29,34	1,968	*Very brisk.*
25	36,67	3,075	
30	44,01	4,429	*High winds.*
35	51,34	6,027	
40	58,68	7,873	*Very high.*
45	66,01	9,963	
50	73,35	12,300	*A storm or tempest.*
60	88,02	17,715	*A great storm.*
80	117,36	31,490	*An hurricane.*
100	146,70	49,200	*An hurricane that tears up trees, carries buildings before it, &c.*
1	2	3	

The Smeaton-Rouse scale of 1759

The outstanding feature of the scale was that the perpendicular pressure of the wind on the mill-sails offered an objective means by which the power of the wind could be calculated and named in an agreed way. It showed, moreover, how an otherwise uncalibrated object such as a windmill might come to be used as an accurate meteorological instrument.

Smeaton's ideas caused a stir within scientific circles and won him the Royal Society gold medal for 1759. The offprinted pamphlet found a wide circulation, going through several editions, and was even translated into French in 1810. This was how the Smeaton-Rouse wind scale, landlocked though it was, came to the attention of Alexander Dalrymple (1737–1808), the first hydrographer to the British Navy. In Dalrymple's view, the Smeaton-Rouse wind scale offered valuable lessons for the improvement of maritime meteorology, and he was determined to see it, or something very much like it, adapted for use at sea. He had, moreover, his own ideas of how this might be achieved.

Dalrymple had devoted his life to traveling over and thinking about the sea, and he became an avid collector of marine paintings and drawings. During his life he had built up a fine collection containing "many specimens of many Masters and Painters of different periods," which he admired as much for their meteorological content as for their stirring naval themes. He believed, as he was later to state in his will, "that it would be very useful to have a series of prints of sea pieces to explain the degrees and gradations of wind from a Calm to a Storm, which Gradations are so well expressed in pictures in my collection," and he went on to bequeath his gallery of paintings to the Admiralty in the hope that they might be of "benefit to the publick."[16] Smeaton's work so furthered Dalrymple's sense of the connectedness of meteorological conditions on land and sea that he drew up a chart showing "the said Gradations of Wind taken from the Sea Journals compared with Mr. Smeaton's Scale from the Work of Windmills," although this document, sadly, has never been found.[17] A draftsman was also reportedly employed on the work of copying the sea pictures for the same purpose, but Dalrymple's sudden dismissal from his post ended all such plans with an unforeseen finality. Nothing is known of the reasons for his dismissal beyond the fact that he had somehow given offense to his superiors. The rumors that later circulated put the affair down to an intemperate "excess of zeal." Dalrymple had always been highly strung, and the shock of his dismissal proved too much for his sensitive mind to bear. He died three weeks later, on June 19, 1808, a humiliated, brokenhearted man.

All his work on the grading of the wind might have been forgotten, had it not been for an updated edition of William Falconer's *Universal Dictionary of the Marine* that was published in 1815, "modernised and much enlarged" by William Burney of Gosport. Burney, one of Luke Howard's warmest supporters, was one of the first meteorological journalists to upgrade his magazine contributions with the new nomenclature of clouds. He was clearly also on the lookout for a new classification of winds, for in his updated version of Falconer's reference book, Burney's entry for "Breeze" reads:

> Mr. Dalrymple, late hydrographer to the Right Honourable the Lords Commissioners of the Admiralty and East India Company, scientifically arranged the wind in the following order:—
>
> 1. Faint Air.
> 2. Light Air.
> 3. Light Breeze.
> 4. Gentle Breeze.
> 5. Fresh Breeze.
> 6. Gentle Gale.
> 7. Moderate Gale.
> 8. Brisk Gale.
> 9. Fresh Gale.
> 10. Strong Gale.
> 11. Hard Gale.
> 12. Storm.[18]

Although hardly the "scientific" arrangement that was claimed, this was nevertheless a clear expression of the same widely felt desire to produce an objective, numbered, ascending scale of wind that had also preoccupied Francis Beaufort. But there was an obvious way for it to be improved, through a comparison of the wind with its effects on an object. This idea had made its way from Rouse to Smeaton, and then from Smeaton to Alexander Dalrymple. The idea was soon to be taken up and refined by the similarly preoccupied—although better connected and clearer thinking—Beaufort.

Dalrymple had become friendly with the ambitious young captain, who was flattered to receive invitations to dine at the Royal Society Club to discuss the state of maritime meteorology in comfortable and conducive surroundings. Dalrymple listened to Beaufort's account of his idea for a scale of the wind and responded with his own description of the much admired Smeaton-Rouse example. He encouraged Beaufort to look at it in greater depth, knowing that the template it offered was just what Beaufort needed for the improvement of the application of his scale.

The idea was a simple one: the value of the Smeaton-Rouse measurement had lain in the objective way by which the force of the wind was determined by the perpendicular pressure it exerted on the sails of a windmill. Beaufort, with Dalrymple's encouragement, adapted the idea to the sails of a seagoing vessel. By means of observations of the typical behavior of a "well-conditioned man-of-war" (just such a ship as the *Woolwich* had once been), the three separate elements of the scale—the number, the common name, and the account of their effect on the ship—could now be combined to create the first objective measurement of the strength of the wind at sea.

It was a near perfect solution to the problem, as Beaufort was instantly aware. He lost no time in rewriting his original fourteen-point scale, as well as slightly reducing it, to read as shown in the figure opposite.

It was a great improvement, achieved through the simplest change of emphasis. Following the example of the windmill scale, the recast Beaufort scale no longer attempted to describe the behavior of an isolated wind but described instead the integrated behavior of the sails of a square-rigged ship. Because ships, like windmills, were comparable in the ways that words were not, it was a brilliant conceptual advance. Thanks to the earlier efforts of Smeaton and Rouse, and the friendly encouragement of Alexander Dalrymple, Beaufort had accomplished his breakthrough.

The new wind scale still had to be accepted, however, and this process was to take far longer than in the case of Howard's classification. In 1829 Beaufort was appointed Hydrographer to the Navy, taking over his mentor Dalrymple's former position, and it was only then that he was able to wield enough influence to begin the task of the official promotion of his scale. Unofficially, that job had already begun, with memoranda sent out to individual commanders of surveying ships. Robert FitzRoy, the commander of the *Beagle*, was one of these recipients and was given firm linguistic instructions by Beaufort concerning the keeping of the logbook: "The state of

FIGURES to denote the Force of the Wind.		
0 Calm.		
1 Light Air, - -	Or just sufficient to give steerage way.	
2 Light Breeze, - -	Or that in which a well-conditioned man-of-war, with all sail set, and clean full, would go in smooth water, from - - - - -	1 to 2 knots.
3 Gentle Breeze, -		3 to 4 knots.
4 Moderate Breeze,		5 to 6 knots.
5 Fresh Breeze, - -	Or that to which she could just carry in chase, full and by	Royals, &c.
6 Strong Breeze, -		Single-reefed topsails and top-gallant sails.
7 Moderate Gale, -		Double-reefed topsails, jib, &c.
8 Fresh gale, - - -		Triple-reefed topsails, &c.
9 Strong Gale, - -		Close-reefed topsails and courses.
10 Whole Gale, - -	Or that with which she could scarcely bear close-reefed maintopsail & reefed foresail.	
11 Storm, - - - - -	Or that which would reduce her to storm stay-sails.	
12 Hurricane, - - -	Or that which no canvas could withstand.	

If the above mode were adopted, the state of the wind might be regularly marked, in a narrow column, on the log-board every hour.

LETTERS to denote the State of the Weather.

b Blue Sky; whether clear or hazy atmosphere.
c Clouds; detached passing clouds.
d Drizzling Rain.
f Foggy—f̣ Thick fog.
g Gloomy dark weather.
h Hail.
l Lightning.
m Misty hazy atmosphere.
o Overcast; or the whole sky covered with thick clouds.
p Passing temporary Showers.
q Squally.
r Rain; continued rain.
s Snow.
t Thunder.
u Ugly threatening appearances.
v Visible clear atmosphere.
w Wet Dew.
· Under any letter, indicates an extraordinary degree.

By the combination of these letters, all the ordinary phenomena of the weather may be expressed with facility and brevity. *Examples:*—Bcm, Blue sky, with passing clouds, and a hazy atmosphere. Gv, Gloomy dark weather, but distant objects remarkably visible. Q̣pdlṭ, Very hard squalls, with passing showers of drizzle, and accompanied by lightning, with very heavy thunder.

The second Beaufort scale

the wind and weather will, of course, be inserted; but some intelligible scale should be assumed, to indicate the force of the former, instead of the ambiguous terms 'fresh', 'moderate', &c., in using which no two people agree; and some concise method should also be employed for expressing the state of the weather. The suggestions contained in the annexed printed paper are recommended for the above purpose."[19] The annexed paper was, of course, the revised Beaufort scale as shown above.

FitzRoy, to his credit, was immediately persuaded of the value of the scale, just as he had also been persuaded to take the young Charles Darwin with him on the same now-legendary trip. Beaufort had been the Admiralty go-between for the two men, and Darwin offered him lavish praise in the preface to his published account of the voyage. Once under way, the 22-year-old Darwin shared a corner of FitzRoy's cabin, from where, along with the increasingly indispensable wind scale, he began the steadfast journey that would take him from Plymouth Harbor to the Galápagos Islands and onward into the popular imagination.

The alliance was not always a happy one. According to Darwin's autobiography, "Fitz-Roy's temper was a most unfortunate one. This was shown not only by passion but by fits of long-continued moroseness against those who had offended him," and more often than not those offenders included Darwin, who recorded numerous onboard quarrels, some "bordering on insanity," that marked their five years together at sea.[20] In spite of the tensions and the tantrums, though, the whole unlikely arrangement on board the *Beagle* went on to constitute, in the words of Beaufort's biographer, "one of the most momentous episodes in scientific and intellectual history," even if FitzRoy, as a militant antievolutionist, was later to regret the alliance altogether.[21] Darwin, for his part, went through life "always afraid of unintentionally offending him," and later complained of FitzRoy's public indignation over the publication of *The Origin of Species*. Beaufort could have had little idea of the consequences of recommending Darwin for the voyage, and still less of the impact of his scale upon the irascible FitzRoy himself. For once back on dry land, FitzRoy's passion for meteorology remained unabated and culminated in his appointment as the first chief of the meteorological department of the Board of Trade, the body that would go on to form the Meteorological Office. His indefatigable energies, when not expended on quarreling with his enemies—even, on occasion, challenging them to duels—were concentrated on issuing weather forecasts. His predictions, however, were every bit as unsuccessful as Lamarck's earlier efforts had been, and his habit of sending them off to no fewer than eight daily newspapers only succeeded in exposing their weaknesses to a wider public. The ridicule he attracted for his failed endeavors greatly displeased the government, which was sensitive to the slightest threat to the reputation of the Board of Trade. FitzRoy was reprimanded, gently enough, but for him it was a humiliation too far. Like Dalrymple before him, the balance of his mind had always been delicate, but the combination of overwork, insufficient resources, and the laughter of the press pushed him rapidly into a depressive decline. He committed suicide with a razor to the throat on April 30, 1865.

Meanwhile, back in his offices in Whitehall, Beaufort and his deputy, Lieutenant Alexander Bridport Becher, continued their campaign on behalf of the scale. They wrote an article for the newly launched *Nautical Magazine* that put the case for clarification with a damning description of the state of the then typical naval logbook:

> "Fresh breezes, and cloudy," in sprawling characters, occupy, with a provoking distinctness, an immensity of space, to the exclusion of some more important remark . . .
>
> The foregoing observations presented themselves, on the consideration of a method for expressing any particulars of the wind and weather, by means of numbers and letters. This method, which originated with Captain Beaufort, the present Hydrographer to the Admiralty, is the result of long experience, and affords a concise means of expressing fully the meaning of whole sentences in writing.[22]

This concern with the content of naval records was a long-term preoccupation of Beaufort's, and it was clear that the war against ambiguity was still being waged on all fronts. Beaufort had also begun to look into further means of unifying the language of the elements as they were used on board ship. To this end, his list of shorthand ciphers was reduced to seventeen letters for describing the state of the weather, with a dot drawn under those to indicate an "extraordinary degree" of the condition, as shown above.[23] Beaufort had used dozens of symbols over the years, including ones denoting individual types of cloud (such as "ci" for cirrus, a code still in use today).

Although a good number of individual ships' commanders were soon employing the Beaufort scale, it was not until 1838 that the Admiralty officially made its use mandatory on board every vessel of the British Navy. Beaufort was convinced that this in itself would produce a valuable resource for meteorological research. Every year a thousand ships each deposited half a dozen logbooks at the Navy Office, but only now were their contents compatible. Beaufort had single-handedly transformed a yearly pile of six thousand testimonials into a comparable source of meteorological information. As he himself was proudly aware, his work on weather, like Luke Howard's before him, had triumphed over ancient ambiguity.

The scale's first use, however, may well have had more to do with discipline than the weather. Because the Beaufort scale describes the behavior of a full-rigged ship, it can also, by extension, be used to ascertain the conduct of its captain and his crew. Numbers 5 to 9 on Beaufort's second scale refer to a well-conditioned man-of-war "in chase"—that is, intent on engaging with an enemy ship. As had Howard with his classification of clouds, Beaufort had devised his scale against a background of war and blockade. This had lent an added impetus to the demands of naval accuracy. A captain who allowed the enemy to outrun him, whether by carrying too much or too little sail, was soon going to find himself in trouble. Too much sail signified an ill-judged enthusiasm, while too little sail showed a lack of appetite for the fray. The former kind of error would mean a reprimand at most, while the latter meant court-martial and the likelihood of instant dismissal for the captain, whose defense in such circumstances had often, up to then, been related to the deflecting power of the wind. But not anymore: Beaufort's wind scale, in describing precisely the state of a ship under sail, and thus the exact power of the prevailing wind, was so useful to naval prosecutors that it might have been expressly designed to offer material evidence to the courts.[24] So as ambiguity was replaced by charts of admissible evidence, meteorological logbooks were transformed into damning courtroom exhibits.

Other wind scales, meanwhile, remained in use in other countries during the following two decades, but when the Beaufort scale was adopted at the first International Meteorological Conference in Brussels in 1853, it brought all such competition to an end. The Brussels conference had been jointly convened by

British and American representatives specifically to discuss standards in maritime meteorology. When Beaufort's scale was proposed by the British representative, Frederick William Beechey, it received unanimous approval from the panel. Later, a meeting of the Permanent Committee of the First International Meteorological Congress in Utrecht voted to adopt the scale for use in international telegrams, thereby fixing it forever as part of the global language of weather. So when Bram Stoker, the author of *Dracula*, described in the novel a wind that blew "from the south-west in the mild degree which in barometrical language is ranked 'No. 2: light breeze,' " he was exploiting the atmospheric possibilities of a recently adopted language.[25]

Like Howard's classification of clouds, the Beaufort scale of wind force has remained in use with only minor changes effected by the World Meteorological Organisation (an intergovernmental agency founded in Geneva in 1951).

Furthermore, the world's very worst winds now have their own dedicated scales, after Antarctic explorers extended the Beaufort calibrations to Force 18 in an effort to accommodate the blizzards that scream around the poles.[26] Tornadoes and hurricanes, too, have the Fujita and Torro scales, which go far beyond the limitations of Beaufort's everyday scale. And even a blizzard registering Force 18 goes well beyond the reach of everyday language and, thankfully, beyond everyday experience.

Wind Speed (Km/hour)	Beaufort Number	Effects at sea	Effects on land	WMO Designation (1964)
Under 1	0	*Sea like mirror*	*Smoke rises vertically*	*Calm*
1–3	1	*Ripples, no crests*	*Vanes do not move*	*Light air*
4–7	2	*Small wavelets*	*Wind felt on face*	*Light breeze*
8–12	3	*Large wavelets*	*Light flags extended*	*Gentle breeze*
13–18	4	*Small waves*	*Small branches move*	*Moderate breeze*
19–24	5	*Moderate waves*	*Small trees sway*	*Fresh breeze*
25–31	6	*Large waves*	*Large branches move*	*Strong breeze*
32–38	7	*White foam from breaking waves*	*Whole trees move*	*Near gale*
39–46	8	*Moderately high waves*	*Twigs broken off*	*Gale*
47–54	9	*High waves*	*Structural damage*	*Strong gale*
55–63	10	*Very high waves*	*Trees uprooted*	*Storm*
64–72	11	*Exceptionally high waves*	*Damage*	*Violent storm*
73 or higher	12	*Air filled with foam and spray*	*Widespread damage*	*Hurricane*

The Beaufort scale today

The first decade of the nineteenth century turned out to be a significant era for meteorological clarity, with the winds following hard on the heels of the clouds as subjects of redefinition. Like Coleridge's painful Lakeland search for a language of aesthetic authority, the character of both the winds and the clouds was approached by an appeal to their powers of affect. Not the wind itself, but its visible force; not the clouds themselves, but their disposition to change: it was these that were to be measured and tracked.

Luke Howard and Francis Beaufort, close in age and in social standing, arrived at comparable solutions to comparable problems, and did so only a few years apart. Neither solution was the first to have been offered, and neither had a particularly complex structure, but both proved best for the purposes of their times, and both have endured into our own. In Admiral Sir Francis Beaufort, F.R.S., the winds had finally found their Howard.

Chapter Eleven

Goethe and Constable

> . . . therefore my winged song thanks
> The man who distinguished cloud from cloud.
>
> *J. W. von Goethe, 1820*[1]

In spite of his deep reservations on the matter, Luke Howard's fame and influence continued to spread across the world as the impact of his names for the clouds moved beyond the immediate limits of his science. This was vividly illustrated one morning in December 1821, shortly after his election to a fellowship of the Royal Society of London, when he received a letter from an unknown clerk employed at the Foreign Office on Downing Street. Fitted out with a parade of flattering terms, the letter begged to inform him that he had an ardent admirer in the greatest literary and scientific figure of the age. Johann Wolfgang von Goethe, no less, had grown impatient to learn some details about Luke Howard's background and personal life. Should he be able to supply some at his earliest convenience, Goethe and his circle of scientific friends would be most flattered and grateful.

Howard was taken aback by this, immediately assuming that the letter was a hoax. Reading it now in its entirety, it is really not hard to see why:

Foreign Office
Downing Street
December 13, 1821

Sir,

Your philosophical labours have attracted so much notice not only in this country, but abroad, that I find myself under the necessity, though an utter stranger, of addressing you.

Mr. de Goethe, one of the ministers of the Grand Duke of Saxe-Weimar, and better known as a poet and philosopher, is so much pleased with your Theory that last summer he published some elegant verses in commendation of it. These were inserted in the original accompanied by a translation, in Gold's London Magazine for July last. I flatter myself they would not displease you. As he takes decidedly the lead in Germany among the poets, I find frequent allusions to your Theory in many other German poems.

By one of yesterday's mails, Goethe, in a letter to me, expresses a great desire to know as much as I may be able to learn about Mr. Luke Howard, thinking no doubt that, from the rank you hold in science, every particular of your life is well known to the learned circles. It may be so; the F.R.S., the F.L.S., etc., might perhaps be able to furnish me with what would fully satisfy Mr. de Goethe. But alas! Sir, I am only a clerk in this office and have no communication with those learned bodies. I am therefore compelled to incur the charge of boldness and importunity by making a direct application to you, and requesting that you would please to favour me with a brief memoir of you.

Mr. de Goethe, though now seventy-two or three, still publishes, and what he writes is read by all his countrymen and women of any literature. If I were so fortunate as to prevail on you to comply with my request, Goethe would doubtless take a pride in introducing your memoir in his periodical publication.

Hoping that under the above circumstances you will pardon the liberty I have taken, I have the honour to be, Sir

Your most obedient and humble servant,

John Chr Huttner[2]

Howard had absolutely no idea what to make of either the letter or its sender. He knew that his essay had been translated for inclusion in French and German periodicals, and he knew that it had caused as much of a stir in Europe as it had in Britain, but he was at a loss to explain the contents of the letter. He had assumed that whatever direct communication he had had with European meteorologists had long since petered out. He had, it was true, once visited Germany, in 1816, but only as part of a group of Quaker philanthropists, distributing the money they had raised on behalf of the victims of the Napoleonic Wars. The trip had had a high profile and was successful in heightening awareness of the atrocities committed by the armies of Napoleon, but it had allowed him little time for furthering his scientific contacts. As far as Howard knew, it was his nomenclature rather than his name that had fixed itself into the German language, and he was more than happy for this to have been the case. It was his clouds, not himself, that he had sought above all to promote. It did not ring remotely true, therefore, that a figure as imposing as Baron von Goethe would suddenly be seeking news of Luke Howard.

Goethe, Europe's greatest intellectual icon, had lived a public life of astonishing, epic achievement. His manifold renown was as a poet, dramatist, novelist, philosopher, traveler, artist, politician, and natural scientist. By the 1820s, when the letter to Howard arrived, he had exercised unrivaled influence over the literate world for close to half a century. His work more than anyone's had lent stature to German literary culture, and though in later years his attentions were divided increasingly between his political and scientific affairs, his intellectual scope remained famous and formidable. His word was enough to settle any matter, and as an arbiter of cultural and civilized values, he was not only (according to Carlyle) the wisest man of the age, but also one of the busiest. It was, therefore, quite understandable for Howard to assume that the letter from "Huttner" was a sham. He viewed it as the prelude to a practical joke, and not a particularly subtle one at that. Perhaps a friend or a member of his family considered fame to have dented his sense of proportion, and that a fraudulent overture from a European giant would lead to an entertainingly awkward confrontation. Whatever or whoever lay behind it, he resolved to turn the tables on what was clearly a mischievous imposture.

But after his inquiries failed to unearth a culprit, Howard approached his oldest friend, William Allen, with the problem of the letter's validity. If anyone could get to the bottom of the story, it was Allen, and he was only too happy to oblige. The approach he took was characteristically direct, and within a matter of days he had come up with the surprising answer:

Stoke Newington 23 of 1 mo 1822

Dear Luke,

Although I hope to see thee here at 12 o'clock on fifth day next and have the pleasure of thy company to Dinner, I thought I would lose no time in informing thee that I saw Huttner yesterday and find that the thing is no Hoax—Goethe one of their very celebrated Poets at Weimar (I think), has a very prodigious inclination to sing the Praises of thy Theory of Clouds—Huttner appears to be his humble Friend and purveyor of News from England—he seems a very respectable man and I am glad in being acquainted with him as he may be useful to us in giving information respecting some parts of the Continent—I do hope that thou wilt call upon him as he is very anxious for it—he is much confined in his office which appears to be rather a subordinate one as Translator in Lord Castelereaghs Office which is on the left hand in the Square in Downing Street—He seemed quite delighted when I gave him reason to hope that thou wouldst call upon him the first time that thou went that way—I think it would be well if thou went to make his Friend a Present through him of thy work on Meteorology—it would set them going in Germany & do good.

I remain dear Luke's
affectionate Friend
W Allen[3]

William Allen had doorstepped Hüttner at his poky office in Downing Street, and there, much to his surprise, he had every detail of the letter confirmed as true. Goethe really was an admirer of Luke Howard, he really had written poetry about him, and he really was anxious to learn the story of his life and times. Goethe's encounter with the classification of clouds, explained "Huttner," had given him enormous pleasure. For some time he had been speaking of little else, and all in all, it seemed as if the old man of letters had been granted a new lease of life.

The claim was no exaggeration. Goethe had always been sensitive to atmospheric effects and interested in their cause and development. The very first entry in his Italian travel diary, for September 3, 1786, saw him waving a glad farewell to the brooding German sky, whose "upper clouds were like streaked wool, the lower heavy. After such a wretched summer, I looked forward to enjoying a fine autumn."[4] The weather had helped precipitate his desire for escape, and when a few days later he asked to be excused for "talking again about temperature and clouds," it was the cold wet summer he had endured in the north, with its clouds "still massed over the mountains of the Tirol," that lay most heavily on his thoughts.[5] The fallout from the eruptions of 1783 was still being felt in the persistently cold weather in the north, and Goethe's Italian journey, undertaken as a temporary respite from his complex network of loves and labors, remained overlaid with the weather anxiety that marked the 1780s. "The weather of the declining year has been so bad," he observed, yet "God does nothing about it."[6] Like so many tourists on their way to the warm South, he was traveling in pursuit of the sun.

What he was surprised to discover, however, was an emerging fascination with clouds. Like John Evelyn in the Alps in 1644, and like Howard himself on the summit of Helvellyn, Goethe was captivated by the spectacle of an evaporating mountain stratus. Unlike John Evelyn, however, he was determined to understand it, and he began to develop his fledging theory of the gravitational influence of the earth on climate. The power of his descriptions, it must be said, was greater than that of his reasoning:

> It clung to the steepest summit, tinted by the afterglow of the setting sun. Slowly, slowly, its edges detached themselves, some fleecy bits were drawn off, lifted high up, and then vanished. Little by little the whole mass disappeared before my eyes, as if it were being spun off from a distaff by an invisible hand.[7]

He excused himself once again for dwelling so long on his "strange" gravitational theories, but he sensed that they were fully borne out by the event he had just described. He claimed that it was the "secret influence" of the mountain ranges rather than any independent atmospheric motion that determined the "elasticity" of the air. The cloud whose death he had silently witnessed had succumbed to an "inner struggle of electrical forces" driven by the "pulse" of the mountains.[8]

Goethe was fond of his curiously geocentric theory of cloud formation, finding it difficult to dislodge; although in the emotional wake of his return to Weimar in 1788 he was to give no further thought to the subject—at least not until the advent of Luke Howard's essay. Most of Goethe's energy had been diverted into other branches of literature and learning, as well as into the complexities of high office. Goethe held the position of privy councillor to Grand Duke Karl August of Saxe-Weimar, a powerful and enlightened patron of the arts and sciences. Under Goethe's tutelage, the duke had developed an abiding interest in British intellectual life, amassing an important collection of English books, prints, and specimens of natural history. He was always on the lookout for the latest scientific news from England.

So when Luke Howard's essay on clouds appeared in Gilbert's German translation in the *Annalen der Physik* in 1815, it was read by the duke and his brilliant teacher, both of whom were instantly captivated. The timing was fortuitous, as Goethe had become increasingly preoccupied with his studies in morphology. *Morphology*, meaning "the science of forms," was a term coined by Goethe to express the connectedness of all natural patterns and formations. In Goethe's view, features such as clouds would have been expected to express the same shaping forces that were responsible for the creation of all other forms in nature, such as whirlpools, snowflakes, or the patterns made by falling leaves. This was fully borne out by Howard's own reasoning, and the essay seemed to Goethe to shed new light on a number of his preoccupations.

Goethe was inspired by the obvious power and simplicity of the conception and claimed that Howard had supplied his morphological studies with a crucial "missing thread."[9] He was dismissive, too, of any kind of science that relied for its conclusions on instrumental data, so he hailed Howard's classification as a moment of "pure" and untrammeled observation: "how much the Classification of the clouds by Howard has pleased me," wrote Goethe, "how much the disproving of the shapeless, the systematic succession of forms of the unlimited, could not but be desired by me, follows from my whole practice in science and art."[10]

Howard's theories of cloud formation thus enhanced the development of Goethe's own view of the "wholeness" of nature, the wholeness of its "mind," as it were, and in his essay "*Wolkengestalt nach Howard*" ("Cloud-shapes According to Howard"), he praised the achievements and evident humanity of the brilliant young English meteorologist.[11] But this was only the beginning. Goethe's admiration and his sense of indebtedness to Howard's meteorological theories did not rest there, but led to one of the most extraordinary personal homages ever paid by one scientific worker to another.

In a series of rhapsodic poetic fragments, eventually to be given the umbrella title "Howards Ehrengedächtnis" ("In Honour of Howard"), Goethe set out to explore the mood as well as the mechanics behind the three major families of cloud, plus the combination cloud then thought of as nimbus. Taking each as the subject and the setting of a stanza, he transformed the opening section of Howard's essay into a sequence of musical passages:

Stratus

When o'er the silent bosom of the sea
The cold mist hangs like a stretch'd canopy;
And the moon, mingling there her shadowy beams,
A spirit, fashioning other spirits seems;
We feel, in moments pure and bright as this,
The joy of innocence, the thrill of bliss.
Then towering up in the darkening mountain's side,
And spreading as it rolls its curtains wide,
It mantles round the mid-way height, and there
It sinks in water-drops, or soars in air.

Cumulus

Still soaring, as if some celestial call
Impell'd it to yon heaven's sublimest hall;
High as the clouds, in pomp and power arrayed,
Enshrined in strength, in majesty displayed;
All the soul's secret thoughts it seems to move,
Beneath it trembles, while it frowns above.

Cirrus

And higher, higher yet the vapors roll:
Triumph is the noblest impulse of the soul!
Then like a lamb whose silvery robes are shed,
The fleecy piles dissolved in dew drops spread;
Or gently waft to the realms of rest,
Find a sweet welcome in the Father's breast.

Nimbus

Now downwards by the world's attraction driven,
That tends to earth, which had upris'n to heaven;
Threatening in the mad thunder-cloud, as when

Fierce legions clash, and vanish from the plain;
Sad destiny of the troubled world! but see,
The mist is now dispersing gloriously:
And language fails us in its vain endeavour—
The spirit mounts above, and lives forever.[12]

For Goethe the identification and naming of the clouds had done nothing less than transfigure mankind's relationship with aerial nature. The clouds had been released into the scientific consciousness, from where they could reach further, into the realm of the pure intellectual spirit, as addressed in the last line of "Nimbus." The greatness of Howard's classification, for Goethe, was that it accounted for the material forces of cloud formation while allowing for the immaterial forces of poetic response to be heard. And his poems, like the essay that preceded them, took the form of just such a response. Art could answer science, it could find within it not only a source of subject matter but a source of real inspiration. Goethe's cloud poems, as reactions to an energizing scientific insight, were heartfelt, joyous, and sincere.

The four cloud poems were first published in German in 1817 in an early number of Goethe's own periodical, *Zur Naturwissenschaft überhaupt*, as the coda to his laudatory essay "Cloud-shapes According to Howard." They were soon attracting the attention of readers, not least that of a young German clerk who worked as a translator in the Foreign Office in London, Johann Christian Hüttner.

Hüttner was a keen admirer of Goethe who had found himself well placed to be of material help to the hero of his nation. His offer to Goethe of any assistance he might need in conducting his English correspondence was gladly taken up. Hüttner soon became the unofficial conduit between London and the Weimar court, sending regular parcels of English books to the increasingly Anglomaniac grand duke, who sent him batches of new German publications in return. In one of these batches was the periodical in which Goethe's cloud poems first appeared. Hüttner was impressed by the poems' beauty and seriousness, and thought that they merited a wider audience. He passed them on to John Bowring, a Foreign Office friend and an experienced translator of poetry, asking whether he might be able to render them into rhyming English verse. Bowring obliged and quickly produced the glowing translations used above.

Hüttner was impatient to see these English versions appear in print, but something about them was bothering him. The verses on their own, he thought, magnificent though they were, seemed overly abstracted, lacking in independent life outside the immediate context of Luke Howard's original essay. When he wrote to Goethe in February 1821 to inform him of his plans to have the cloud poems published in English, Hüttner took the opportunity to suggest that a few introductory lines might serve to place the personal celebratory nature of the fragments in a clearer context. Perhaps, to this end, he might add something at the start about Howard? Goethe, long habituated to composing on demand, immediately complied with the request, and only a few days later, his young friend in London was reading with delight the following introductory stanzas:

> When Camarupa, wavering on high,
> Lightly and slowly travels o'er the sky,
> Now closely draws her veil, now spreads it wide,
> And joys to see the changing figures glide,
> Now firmly stands, now like a vision flies,
> We pause in wonder, and mistrust our eyes.
>
> Then boldly stirs imagination's power,
> And shapes there formless masses of the hour;
> Here lions threat, there elephants will range,
> And camel-necks to vapoury dragons change;
> An army moves, but not in victory proud,
> Its might is broken on a rock of cloud;
> E'en the cloud messenger in air expires,
> Ere reach'd the distance fancy yet desires.

But Howard gives us with his clearer mind
The gain of lessons new to all mankind;
That which no hand can reach, no hand can clasp,
He first has gain'd, first held with mental grasp.
Defin'd the doubtful, fix'd its limit-line,
And named it fitly.—Be the honour thine!
As clouds ascend, are folded, scatter, fall,
Let the world think of thee who taught it all.[13]

Read now as an ensemble, the seven verses are both a celebration and an invocation of Howard's classification of clouds. The "mental grasp" of the imagination, only hinted at in the original four stanzas, is now directly named as the power behind Howard's enhanced understanding of the clouds, and the warmth with which the Englishman himself is personally commended—"Be the honour thine!"—elevates the poem to the status of a lasting scientific tribute.

Hüttner, who had now assembled the complete poem that he wanted to see published, passed the three additional verses on to George Soane, a writer who had recently distinguished himself with an acclaimed English version of Goethe's *Faust*. As soon as Soane had completed his part of the translation, Hüttner arranged for the whole thing to appear in both English and German in the July 1821 issue of

Gold and Northhouse's *London Magazine*, with an additional interpretative commentary on the poems written almost certainly by the energetic Hüttner himself:

> These three first strophes have not hitherto appeared in print, and it is from a strange chance only that they ever fell into our hands.—Goethe having observed that something was really wanting to his poem, in honour of the celebrated Howard, to make it finished and intelligible, resolved to write three strophes as an introduction.
>
> In the first strophe, the Indian Divinity, Camarupa (wearer of shapes at will), is introduced as the spiritual being, who shows herself here also active in changing the images according to her fancy, and shapes and unshapes the clouds . . .
>
> In the third strophe, that nothing may be wanting, Howard's name is mentioned, and his merit acknowledged in having fixed a nomenclature thro' which we may adhere in the division and description of atmospheric phenomena.
>
> This very nomenclature is intimated and announced in the penultimate line of the introduction as follows:
>
> As clouds ascend,   Are folded,   Scatter,   Fall.
> Stratus.   Cumulus.   Cirrus.   Nimbus.[14]

As a summation of Luke Howard and his naming of the clouds, the final couplet of the introductory section—"As clouds ascend, are folded, scatter, fall,/Let the world think of thee who taught it all"—is a resourceful stroke, and one which leads naturally into the programmatic evocation of the clouds themselves, as originally conceived by Goethe. In appealing to Goethe for a buildup to the visionary cloudbursts, Hüttner proved that his editorial instincts were correct, for on its second, extended appearance under Hüttner's guiding patronage, "Howards Ehrengedächtnis" became a major work of scientific art.

Although the cloud-by-cloud structure of Goethe's composition was ingenious, it was not in fact an isolated example of the form. It bears a striking similarity to a near-contemporary work written by Percy Bysshe Shelley, almost certainly while living in Pisa with his second wife, Mary Shelley, in early 1820. "The Cloud," while delivering an atmospheric meditation on human creativity, offers a vivid poetic primer for Howard's classification of clouds. The named modifications, with the one exception of cumulostratus, appear united in a fluid, changeable organism that addresses the reader in an appealingly mocking tone of voice:

> I bring fresh showers for the thirsting flowers,
> From the seas and the streams;

I bear light shade for the leaves when laid
In their noon-day dreams.
From my wings are shaken the dews that waken
The sweet birds every one,
When rocked to rest on their mother's breast,
As she dances about the sun.
I wield the flail of the lashing hail,
And whiten the green plains under,
And then again I dissolve it in rain,
And laugh as I pass in thunder.[15]

This first stanza has been interpreted by meteorologists as an accurate invocation of cumulus, stratus, and nimbus clouds, and the poem goes on to characterize cirrus (the moon "glides glimmering o'er my fleece-like floor"), cirrocumulus ("when I widen the vent in my wind-built tent"), and cirrostratus clouds ("I bind the sun's throne with a burning zone, and the moon's with a girdle of pearl").[16] All the modifications were offered up by Shelley as aspects of a single cloud personality (the mocking, mythopoeic "I" who laughingly narrates the poem) in a direct and knowing tribute to Howard's contention that clouds unite, pass into one another, and disperse in distinct and recognizable stages. Shelley's perpetual cloud stands for all clouds in all formations, and its joyous declaration that

I am the daughter of earth and water,
And the nursling of the sky;
I pass through the pores of the ocean and shores;
I change, but I cannot die

is a testament to the clouds' replenishing role as agents of the water cycle. This had long been considered one of nature's most harmonious and philosophically attractive aspects—all that is taken away shall one day be returned—and had been much loved by poets around the world since antiquity. In fact, one of Shelley's sources for "The Cloud" was a recent translation of a celebrated Sanskrit poem, the *Mégha Dúta*, or *Cloud Messenger*, composed sometime in the late fourth century by the great Hindu poet Kālidāsa, a figure known among the Romantic poets as "the Shakespeare of India."

The ancient poem narrates the misfortunes of a servant of the Hindu god of wealth, who is exiled to a mountainside for displeasing his master. Watching the clouds gathering in the south from his lonely vantage point, the servant is moved to ask one of these "wearers of shapes at will" to waft his sorrows down to the wife whom he left behind, far away on the verdant plains. The exile imagines the journey made by the cloud as it moves northward over the continent on its errand of mercy:

On *Naga Nadi's* banks thy waters shed,
And raise the feeble jasmin's languid head:
Grant for a while thy interposing shroud,
To where those damsels woo the friendly Cloud;
As while the garland's flowery stores they seek,
The scorching sun-beams singe the tender cheek,
The ear-hung lotus fades, and vain they chase,
Fatigued and faint, the drops that dew the face.[17]

The cloud, having rained down the message of love on the exile's wife, must then make its return to the mountains to replenish itself through the familiar agencies of convection and condensation.

Shelley's cloud thus mingled recent scientific innovation with ancient poetic tradition, and when it was translated into German by an anonymous hand in 1830, Goethe was enormously delighted.[18]

The promise of Howard's classification of clouds had been quickly grasped by these poets of weather, who had sensed for themselves that nature was governed not by a fixed but by a "mobile order."[19] They realized that the idea of mutability, of disposition to change, fed both the imagination and the understanding, and to have this confirmed on an empirical level came almost as a form of liberation.

Shelley had earlier taken clouds to be his main image of mutability, in the 1814 version of his poem of that name, "Mutability":

> We are as clouds that veil the midnight moon;
> How restlessly they speed, and gleam, and quiver,
> Streaking the darkness radiantly!—yet soon
> Night closes round, and they are lost forever.

Night falls, convection diminishes, and the clouds begin to disperse; but now, thanks to Howard, clouds were no longer mere images of loss. They could, as Wordsworth expressed it, be fixed in glorious shape.[20]

Goethe felt enormously indebted to "the man who distinguished cloud from cloud," and he was keen to offer him the extended hand of friendship and admiration. He wanted to know more about the personal circumstances of this insightful English meteorologist.

According to Goethe's own account, Hüttner's mediation was sought once more, and he was given the task of uncovering any biographical details that might have been available in the public domain:

> I requested a friend always active & obliging—Mr Hüttner of London—to procure for me if possible even the simplest outline of

> Howard's life, that I might comprehend how such an intellect had been trained—what opportunities and what circumstances had brought him to inspect nature so naturally, to devote himself to her—to understand her laws, & again with so great simplicity to unfold those laws in writing. My stanzas to the honour of Howard had been translated in England, being particularly prefaced by an explanatory introduction in rhyme—they were known through the press, & therefore I had reason to hope that some friend would heed my wishes.[21]

Goethe's wishes were quickly heeded, and this again was due to Hüttner's fortunate presence in London. Without him, as Goethe well appreciated, the extended version and translation of the cloud poems would never have materialized. Without Hüttner, we would also know a great deal less about Luke Howard's life and opinions, for it was only in response to the Goethean command that he set about writing his memoirs.

Following William Allen's reassurances that the letter from Hüttner was genuine, Howard set about answering the request for information. He prepared what he described as a summary of "the person who wrote the 'Essay on the Modifications of Clouds' " and sent it directly to the clerk.[22] Hüttner then posted it to Goethe, along with Howard's gift of a copy of *The Climate of London*. The gift of the book was Allen's idea.

The memoir, though brief, contains the only firsthand account of Luke Howard's life, charting the entire development of his scientific, religious, and familial occupations. "I am a man of domestic habits and very happy in my family and a few friends, whose company I quit with reluctance to join other circles," he confessed, the loving warmth of his character shining through the closely handwritten pages as they were read by the poet in Weimar.[23] "Am I a fool for this in Goethe's esteem?" But Goethe, far from viewing Howard as a timid provincial, was greatly moved to discover that the celebrated classifier of clouds, whose work had attracted such poetry and praise, was modest, sincere, and humorous in his outlook. "With but a few words in haste," he wrote back to Hüttner, "I am letting you know that for a long time nothing has given me so much pleasure as the autobiography of Mr. Howard, which I received yesterday and have been thinking of ever since. In truth nothing more pleasant could have happened to me than to see the tender religious soul of such an excellent man opened out to me in such a way that he has been able to lay bare for me the story of his destiny and development as well as his innermost convictions."[24]

Howard's affecting remembrance of his background and early intellectual growth must have reminded Goethe of his own Wilhelm Meister, his last great fictional hero, whose early education and apprenticeship so shaped the outcome of his life and adventures. In Howard's case the schoolboy Latin, the weather events of the 1780s, the exile and return from a dull apprenticeship, and then his immersion in the burgeoning scientific culture of London in the following decades were the shaping contexts of his observations of the clouds. These in turn found full expression in the Askesian Society lecture that was delivered in December 1802.

Goethe was enchanted by the trajectory the memoir revealed. It seemed to confirm as much about the shaping of the scientific personality as the essay had confirmed about the clouds. "There is perhaps no finer example of the sort of mind to which nature delights to reveal herself," he wrote, with a wistfulness approaching the envy of age for the brilliance of his own vanished youth; "with what spirit she condescends to maintain a continual inward communion."[25] For Howard had been chosen by a benevolent nature, seconded by an enthusiastic Goethe, who considered that the noblest achievement of the human mind was "to have probed what is knowable and quietly to revere what is unknowable."[26] This, he believed, was what Howard had achieved both in his classification of clouds and in the pages of his tender self-portrait.

Embraced by Goethe as a kindred spirit, Luke Howard became the only Englishman whom Goethe ever addressed as Master; and the following lines from Goethe's diary form a brief pen portrait of a newly awakened self:

> Disciple of Howard, strangely
> You look around and above you every morning

To see whether the mist falls or rises
And what clouds are showing.[27]

Goethe's interest in clouds, once reawakened, continued unabated for the rest of his life, as did his enthusiasm for the young English author. He promoted the cloud classification throughout the German-speaking world, recommending it particularly to the artists of his acquaintance. The 15-year-old student painter Friedrich Preller, for example, was given a translation of the treatise by Goethe and told to "read that and then observe the various cloud formations and bring me clear drawings of them."[28] Preller, as far as is known, willingly complied with the request. The Dresden-based artists Carl Gustav Carus and Johan Christian Dahl received similar advice from Goethe, with Carus, as a result, experiencing the same level of awakening as had the poet. As soon as he had read Howard's treatise, he too felt that his "problem of how to reconcile scientific analysis with creative freedom had now been solved."[29] Clouds, according to Howard's system, were free to go about their endless reinventions. Goethe won many such converts to Howard's scientific mutability, and German culture as a whole, as Hüttner later reported, was soon reverberating loudly to the names of the clouds.

But not everyone in Goethe's circle was so ready to be convinced about the clouds. The dark and unbiddable Caspar David Friedrich refused Goethe's request to supply him with a set of illustrations for the 1817 essay on Howard, on the grounds that the project would help "undermine the whole foundation of landscape painting."[30] Friedrich resisted any attempt "to force the free and airy clouds into a rigid order and classification," holding firmly to the notion that the deep obscurity and impression of the clouds were valuable attributes in themselves.[31] Clouds, for Friedrich, flew as emblems of a limitless freedom, and the idea of seeking to patrol their boundaries with what he saw as an imposed scientific order filled him with an artist's despair.

Criticism from such an influential source did nothing to discourage the elderly Goethe, who gave as little credence to the notion of the purely aesthetic as he did to the idea of the purely scientific. Although he and Friedrich had been friendly up to then, they were never to see eye to eye again, for the latter's elevation of the poetic imagination above the scientific outlook, like Keats's despairing view of Newton and the rainbow in "Lamia," overlooked the role that imagination has always played in the creation of scientific culture. Goethe was adamant in seeing the two as inseparable, complementary aspects of human consciousness. Art and science, after all, were both products of the human imagination; both were ways of representing and giving order to the world. His comment, for example, on the topographical engravings of David Read, that the works had "something rough about them that displeases one, particularly in the clouds; he may not have sufficiently studied them according to Howard," is a testament to his outlook on painting as an integral aspect of enlightenment.[32] Since the painting of nature is a means toward the clearer understanding of nature, it has a serious, even a profound, job to do.

Howard's ideas had become Goethe's touchstone for the understanding and representation of clouds, not merely in terms of what might be called accuracy, but in terms of what might be called spirit. This was what the influential Victorian critic John Ruskin was later to call the Truth of Clouds, by which he meant a kind of renewed spiritual covenant between mankind, the natural world, and the realm of the divine. Clouds for Ruskin offered a challenge to both the intellect and the soul, challenges to which Howard, in his view, had more than risen. Like Goethe, Ruskin was to grow increasingly preoccupied with thinking about the atmosphere, devoting several chapters of his study *Modern Painters* to lengthy discussions of the sections of the sky, including "the Open Sky," "the Region of the Cirrus," "the Central Cloud Region," and "the Region of the Rain-Cloud." By 1856 he had come to the often repeated conclusion that "if a general and characteristic name were needed for modern landscape art, none better could be invented than 'the service of clouds.' "[33] As a judgment it was worthy of Baudelaire; well, almost.

Yet between the twin critical peaks of Goethe and Ruskin came an even greater exponent of the overarching truth of Howard's clouds: the English landscape painter

John Constable, whose studies of the Hampstead sky have now taken their place among the most revered expressions of the European romantic vision.

Constable's clouds have generated a vast literature as well as a number of major exhibitions, the most recent of which were held in Liverpool and Edinburgh in the summer of 2000.[34] "You can never be nubilous," as Constable once wrote in a letter to a sympathetic friend, for "I am the man of clouds."[35] And judging by the skies that dominate his work, it is difficult not to agree.

"Painting is a science and should be pursued as an inquiry into the laws of nature," Constable claimed in a lecture at the Royal Institution. "Why, then, may not landscape painting be considered a branch of natural philosophy, of which pictures are but the experiments?"[36] Constable asked this question in all sincerity, and he sought throughout his career to provide a satisfactory visual answer. The results of his inquiries, in the form of his studies of clouds and skies, demonstrate the success of his experimental approach to painterly creativity.

Constable was one of the first European artists to paint for fame rather than for money. His thoughts were fixed on posterity, particularly during the summer months of 1821 and 1822, which he spent on the higher slopes of Hampstead Heath, working with an anxious intensity that few artists had ever achieved before. His concentration was focused on the extension of his observational range, and clouds were the means he had chosen for the task. After years of searching for an

John Constable, Study of Clouds at Hampstead, September 1821

isolated image, seeking a motif upon which to weigh his technical advancement as a painter, he had found it at last in the unending sequences of clouds that emerged and dissolved before his eyes like images on a photographic plate. Like Howard, Constable was struck by the effortlessness of their silent transformations. The sky became the true Constable Country, and the series of more than a hundred cloud studies that he made in those two brief summers are among the most admired of all his productions. Not as boldly iconic as *The Hay Wain*, perhaps, nor as amenable for the purposes of patriotism or nostalgia, Constable's cloud paintings can nevertheless be counted among his most moving and mysterious works. He described them collectively as "noble clouds & effects of light," and he was right to stress the clouds' nobility, for they are carried aloft by a weightless magnificence that is all their own. His paintings are the enormous gestures of the sky writ small.

Each summer morning Constable walked the short distance from his lodgings at Lower Terrace, at the southern end of Hampstead village, to the meadows of Prospect Walk, his favored spot for viewing the great scenery of the passing sky. In returning to the same locations at the same times each day, in order to build up a measured picture of the life of the skies over time, his methods shared as much with the discipline of fieldwork as with the demands of painting from nature. Constable was not the first to have painted from life on Hampstead Heath; neither was he unique in having chosen clouds as the subjects of his regular trials. But he was among the first to use the discoveries of a developing science to pursue a new direction in his craft. The methods of an old art, painting, and a new science, meteorology, were combined in a single unified outlook. This outlook was expressed in the claim he maintained, almost as an item of faith, that "we see nothing truly till we understand it." There would be none of Keats's or Friedrich's mystification in Constable's meteorology. This was to be a serious engagement with the laws of aerial nature.

Constable's experiments in pursuit of meteorological understanding were conducted not by means of a recording barometer but in the brilliant crucible of paint. The scene around his folding seat was like an outdoor manufacturing laboratory. His painting box was filled with a litter of alchemical vials, whose labels, RED OXIDE OF MANGANESE, PROTIODIDE OF MERCURY, and SESQUIOXIDE OF CHROMIUM, recalled the basement world of the Plough Court laboratory, where artists' materials were made to order amid the pharmaceutical apparatus. But if such equipment seemed the outward sign of his commitment to empirical research, the expression of his results was nevertheless limited by the physical properties of the paint itself. Even thinned down as much as it was, it still could not fly unimpeded across the paper as he wished. For the aspects of the sky changed so fast; high above the Heath, high above the world, unmediated nature migrated into a series of unmediated images whose shadow, tint, and form refused to obey the principles of slow, picturesque composition. These were not the familiar landscapes that were built up over periods of history or stasis. Constable, as had Howard before him, struggled to comprehend the variety of forms, and what he found himself painting was an unresolved process that made unprecedented demands on his powers of concentration. Time itself became a feature of each image, as the busy skies relinquished themselves of their short-lived, transitory forms. The challenge was to be faithful, on both artistic and scientific grounds, to the vision of a particular formation; or rather, the challenge was to be faithful to the *process* of its formation, to put a creative act in colloquy with the creative powers of nature.

Painters before Constable had often approached the problem of weather and clouds, many with great success. With encouragement from a number of artist's manuals, such as those written by Roger de Piles and P. H. de Valenciennes, they were counseled to study the ever changing aspect of the atmosphere. Charles Taylor, a prolific and entrepreneurial engraver, publisher, and critic, even compiled a list of extreme effects which might occupy the attentions of artists:

> The Aurora Borealis, and other lights. And why not the Eclipse? Also Fogs, Mists, and other Exhalations . . . I should like to see a competent idea of a Volcano, near, and remote; of a Hurricane in the West Indies, as distinct from an ordinary storm; of a Waterspout, accurately represented; of a Typhon (Tuffoon) in the Japanese

> Seas; of the Samiel or Purple Hot Wind of Arabia; of the Whirlpool, called the Maelstrom, on the coast of Norway; and of many other curious phaenomena, which introduced into correspondent and accurate Landscapes, would furnish triumphs for the imitative Arts.[37]

Taylor's call was for a catalog of killer weather, no less, with which to amend forever the face of British landscape painting. This was to be the direction taken by Turner, particularly during the late 1810s, when a series of volcanic eruptions around the world once again painted the skies of Europe with a vivid palette of colors. The 1815 eruption of the Indonesian island of Tambora, in particular, set off a chain of brilliant sunsets around the world that were tracked over the Thames in London by a fortunate and captivated Turner.

On a milder level of weather effects, Alexander Cozens, who had been a drawing master at Eton in the 1780s, devised a visual system for teaching cloudscapes. It was a kind of register or classification for aiding his pupils' compositions. The terms of his twentyfold classification, which included "Streaky Clouds at the top of the Sky," "Streaky Clouds at the bottom of the Sky," "Half Cloud half plain," "All cloudy, except one large opening," and so on, recall earlier attempts to organize the clouds by their outward manifestations.[38] Like Robert

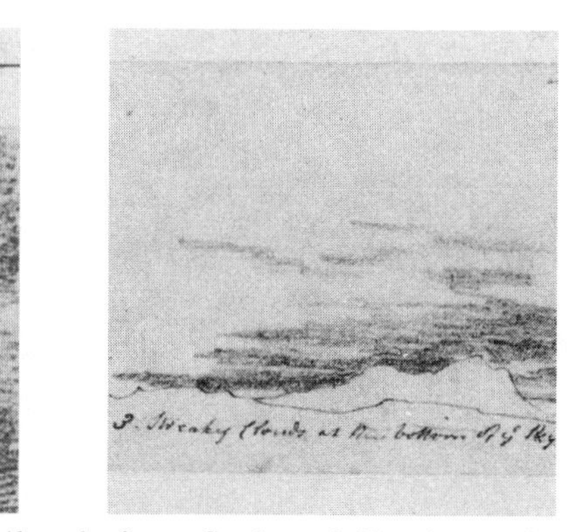

Alexander Cozens, Studies of Clouds, *c. 1785*

Hooke or the observers at the Royal Society of Mannheim, Cozens devised a system for grading the appearances of the sky, albeit with a painterly rather than a meteorological motive. Like Friedrich, Cozens was more interested in light than in Linnaeus, whereas Constable once described heaven as a place where one might walk "arm in arm with Milton & Linnaeus," an image of the marriage of taxonomy and verse.[39] Constable's reverence extended to the entire Linnaean system of classification as adopted by Howard for the clouds, and by the time he had found and copied out the sequence of Cozens's illustrations in 1823, he had long been familiar with the meteorological theories of both Howard and Thomas Forster.

Most of Constable's sky sketches featured weather notes recorded on the back, written in the hours after they had had time to dry. The majority of these notes were descriptively simple ("Showery Wind N. Easterly," for example), but a number were detailed enough to suggest a knowledge gained beyond his own solitary observations, or from a childhood spent working at his father's windmill in Suffolk.[40] And sure enough, among the copiously annotated volumes in his personal library was a copy of the second edition of Thomas Forster's *Researches about Atmospheric Phænomena.* The book was bought secondhand for six shillings, marked by the bookseller as "Published at 10/6 scarce," and as John E. Thornes has shown in his recent book on the subject, Constable made a series of underlinings and annotations on the pages of the opening chapter, "Of Mr. Howard's Theory of the Origin and Modifications of Clouds." As Thornes points out, these comments and underlinings reveal Constable's meteorological knowledge to have been "considerable"; he was able to pick out ambiguities within the cloud classification as reproduced in 1815 and to dispute a number of Forster's proffered conclusions.[41] He certainly took the subject seriously: "This is not correct," he scrawled alongside Forster's account of the generation of rain on page 24; "electrical fluid will convert an [?] without an [?]," the illegible handwriting doing nothing to conceal the confidence of his judgments.[42] Indeed, when Constable later wrote to a friend offering to lend him some books on the subject of "clouds and skies," he made the insightfully disparaging comment that "Forster's is the best book—he is far from right, still he has the merit of breaking much ground," a conclusion that might well have secretly pleased Luke Howard, who by then was having to defend his terms against the incursions of Forster's translations.[43]

In 1836 Constable began to deliver a series of subscription lectures on the history of landscape painting at the Royal Institution on Albermarle Street. The location was an appropriate one, given his overriding interest in scientific subjects, and his first lecture began with the stated conviction that his profession as a painter could be shown to be "*scientific* as well as *poetic*; that imagination alone never did, and never can, produce works that are to stand by a comparison with *realities*."[44] Constable's painting was not art for art's sake: it relied upon nature for its measure.

These public statements in the lecture hall were to prove the summary of his life's interests and achievements, and his intention was to deliver, as the highlight of the series, a lecture on the new science of meteorology. This lecture, however, proved more elusive than most of his ambitions. "My observations on clouds and skies are on scraps and bits of paper," he wrote to a friend in December 1836, "and I have never yet put them together so as to form a lecture, which I shall do, and probably deliver at Hampstead next summer." But by that next summer, Constable was dead, having suffered from depression and ill health for some time. Apart from a brief suggestion of a reading list, in which he made his barbed recommendations of the work of Thomas Forster, no trace of any of his lecture preparations has been found. John Constable on "The Science of Painting the Sky": one of the great lost lectures of the early nineteenth century.

What is clear is that Constable was preoccupied by meteorology mainly as a means of approaching his art. His claim that it would be "difficult to name a class of landscape in which the sky is not the 'key note,' the 'standard of Scale,' and the chief 'Organ of Sentiment,' " imagined the sky as a page filled with metaphors of light. Like Goethe, Constable was attracted to Howard's vision of aerial nature as an endlessly recursive system of signs. Yet, again like Goethe, he was as much preoccupied by the science of clouds and weather, with "the natural history, if the expression may be used, of the skies." Rarely had a visual artist undertaken such serious secondary research as Constable did on the clouds. When Ruskin made his later demand that "every class of rock, earth, and cloud, must be known by the painter, with geologic and meteorologic accuracy," he was describing the outlook pioneered by Constable a whole generation before.

It is this very "meteorologic accuracy," of course, that has given fuel to Constable's detractors. Henry Fuseli once complained that the landscapes of Constable made him want to call for his overcoat and his umbrella.[45] This was more than an image of parochial conservatism, for every canvas in Constable Country carries a threat of oncoming rain. But if he didn't much care for the pictures themselves, Fuseli, the sarcastic Swiss drawing master at the Royal Academy, at least recognized that they had been painted from under the same named clouds that by then had come to dominate a broad cross section of romantic art and science. Luke Howard's work, revered by Goethe in Weimar, by Shelley in Pisa, and by Constable on Hampstead Heath, was now at the epicenter of European culture.

For it was Howard who was "nubilous," in Constable's phrase; it was Howard who was the man of the clouds.

Chapter Twelve

The International Year of Clouds

What is the end of fame? 'tis but to fill
A certain portion of uncertain paper:
Some liken it to climbing up a hill,
Whose summit, like all hills, is lost in vapour.

Lord Byron, 1819[1]

Had they leafed through the back pages of Alexander Tilloch's *Philosophical Magazine* for September 1823, his readers would have noticed, amid Intelligence and Miscellaneous Articles, the following brief announcement:

> Proposed Establishment of a Meteorological Society. The science of Meteorology, we understand, is likely to receive, in a short time, the powerful aid of a Society expressly devoted to its cultivation. A meeting will be held on the third Wednesday in October, at

> the London Coffee-house, Ludgate Hill, at 8 o'clock in the evening, for the purpose of taking the subject into consideration, at which a number of scientific gentlemen, attached to the science, are expected to attend: and we hope their example will be followed by all who are interested in Meteorological pursuits.[2]

In the next issue of the magazine, Tilloch, writing in his best editorial tone of voice, expressed "much satisfaction in announcing the formation of 'The Meteorological Society of London,' " and he printed in full the eleven resolutions agreed on by the attendees at the inaugural coffeehouse meeting.[3] These resolutions included the election of various officers to sit on the council, the election of a body to oversee the printing of the society's proceedings, the election of a body devoted to soliciting meteorological correspondence from around the world, and the collection of an annual subscription fee of two guineas, "to be paid in advance by every member." The last-mentioned item, although agreed to by a vote, was soon to be the source of bitter conflict.

Among the names of the dozen or so founding members, two will by now be familiar: Luke Howard and Thomas Forster. The pair had settled their earlier differences over the English cloud translations and were ready to cooperate on finding the solution to another issue that regularly attracted meteorological complaint: Why was there no society or organization specifically devoted to its study?

The first two meetings of the new society were taken up with administrative issues, dealing particularly with aspects of recruitment. Setting about building up a list of members must have recalled the early days of the Askesian Society, when William Allen and W. H. Pepys were both persuaded to join the group so as to boost the number of familiar faces. Howard's old friend and fellow Quaker George Birkbeck, the founder of the newly opened Mechanics' Institution, was elected to the presidential chair, while Henry Clutterbuck was made treasurer and Thomas Wilford was given the role of secretary. An eight-member council was also established, to which Howard and Forster were appointed. Clearly, no effort was spared in what was intended to be the inauguration of a respectable and permanent institution.

These various arrangements took some time to go through, and the first actual paper to be read at the society was not heard until January 1824, when John Gough's observations on the "Natural History and probable causes of the Vernal Winds of the North of England" were read aloud to an enthusiastic audience. The next meeting, held on February 11, heard Luke Howard speak on the subject "Curious Effects of the Radiation of Heat," ideas which he later incorporated into the second edition of *The Climate of London.*[4]

So far so good, the verdict must have been, at least in the initial months of the society. A forward step had at last been taken in the professionalization of the science of the atmosphere. This was something for which many people had been agitating for years. But the active life of this first society was to prove unexpectedly short. After breaking for the summer in May 1824, the members went their separate ways and failed to hold another meeting for twelve and a half years. The few letters that they wrote to one another during that period of separation show a marked inclination for squabbling over money and resources. Forster complained that "the founding was too small for the necessary spending," while the president, George Birkbeck, grew increasingly despondent over the "want of zeal" that had sunk the young society into such premature decline.[5]

Luke Howard was mostly to blame. As part of a project of withdrawal from London life, he had just bought property near Ackworth, in Yorkshire, where an increasing amount of his time was now being spent. His mother's family was from nearby Pontefract, and he had always loved his visits there as a child. Now that he was both famous and financially secure, the idea of rural retirement in Yorkshire had become an increasingly attractive one.

He was soon busy with his new domestic arrangements, as well as with his hours of voluntary teaching and administration at the nearby Quaker school, the school his father had helped found. He had also become much engaged in other social and philanthropic projects, such as the antislavery campaign and the fund for the relief of the European victims of war. As a consequence he found himself less involved in meteorology than at any previous time in his life. The daily weather readings taken both at home and at work were now in the hands of his family and staff, and with the exception of the second edition of *The Climate of London*, he was to publish no new scientific work for nearly twenty years, occupying himself instead with a series of arcane theological disputes that erupted within the Society of Friends. It seemed almost as if Luke Howard the meteorologist was making an attempt to disappear.

So, naturally enough, his enthusiasm for the corporate side of the science was low. From the start he had found the atmosphere and the spiraling financial problems of the Meteorological Society too dispiriting to face up to, even on a once-a-month basis. Within a few weeks he had just stopped going, pleading the responsibilities of Ackworth as his excuse. And without Luke Howard as its guiding spirit, the circle lacked the energy to continue. The group dynamic, in spite of the promise of its first few meetings, had been as nothing compared with the early fire of the Askesian Society, and that had been nearly thirty years before. The spark of youth had long been extinguished. Most of the old friends were burdened with professional and family commitments, and they, like Howard, were finding it hard to spare the time for talking to their friends about the weather.

What was needed was a younger generation of meteorological researchers, with both the time and the enthusiasm to devote to evenings of organized science. And in November 1836 the Meteorological Society of London did indeed re-form, with a cast of new and younger faces in the membership.[6] By then, much of the original grouping of 1823 had drifted away, unimpressed by the benefits of the monthly meetings, and generally unhopeful about the life expectancy of such a poorly funded and fragmented operation. Those first meetings in 1823 had been held not in a headquarters but in a series of hired rooms, and there simply had not been enough money to support the wider aims of the venture. Indeed, everybody involved in it had been left out of pocket, to their great and lasting irritation.

When the new organization convened in 1836, it reelected Howard as an honorary member, but he refused any further involvement. His sole contribution to party funds was a parcel containing 150 copies of a reprint of his essay on clouds. Like his fellow founder Forster, who by then was living in France, he failed to turn up at any further meetings. The new body, however, flourished without them, despite the many ups and downs in its fortunes. In 1850 it re-formed once more, as the British Meteorological Society, which in turn went on to become the Royal Meteorological Society, in 1883.

In the meantime, during the long interregnum one or two of the original members had died, and for Howard in particular, it was the death of Alexander

Tilloch, one of his oldest acquaintances and also one of the strongest external supporters of the original Meteorological Society, which fell as the bitterest blow.

Tilloch died at his home in Islington on January 26, 1825, following a short, unidentified illness. Most of his life's ambitions had been achieved in the course of his 64 years, apart from his failure to patent a design for the first unforgeable banknote. He had exercised a profound influence on scientific publishing in Britain, transforming its fortunes by launching a journal—the *Philosophical Magazine*—that would outlive him for another two centuries, and that flourishes still to this day.

Tilloch was also a well-loved father to his daughter, whom he brought up alone after the death of his wife, Elizabeth, in childbirth in 1783. His daughter, named Elizabeth after the mother she never knew, went on to make a happy marriage with the Scottish novelist John Galt, who honored the memory of his philosophical father-in-law by putting him into one of his novels. There, Mr. Ascomy, "a philosopher of no ordinary calibre," is given a walk-on part as the father-in-law of the restless hero of *Bogle Corbet*, Galt's first-person narrative of the sorrows of Scottish emigration.[7] There was real affection in the characterization, as well as a skillfully evoked sense of Tilloch's unique conversational style: caustic, allusive, and driven by similes. Anyone who knew him—and Howard knew him well—would have recognized the cameo immediately.

Following a leisurely probate process, a sale of Tilloch's effects was held in London at the Great Room ("The Poets' Gallery") at 39 Fleet Street, on May 16 and 17, 1825, commencing, according to the sale catalog, at "12.30 precisely."[8] The two-day event witnessed the dispersal of an extraordinary life's collection. Alongside paintings, drawings, prints, coins, earthenware, silverware, and the usual congregation of household collectibles, an astonishing array of mathematical and philosophical instruments was sold to an acquisitive audience made up of friends and London dealers. Among the items of the Tilloch hoard were a 15-inch concave mirror in a black frame; a 10-inch-diameter plano-convex burning-glass; a two-and-a-half-foot achromatic telescope complete with brass stand and eyepieces; a Culpepper's microscope in a wainscot box; a "RAMSDEN'S OPTICAL, by JONES"; a small microscope in a fish-skin case; a self-registering horizontal day-and-night thermometer; a copper air fountain with stopcock, pipe, and jet; an iron chemical furnace with tubes and a conducting pipe; a blunderbuss with two brass rings; a Davy's safety lamp; twelve globe receivers with long necks; three boxes of minerals; about 2 pounds of quicksilver in a bottle; an elliptical tracing frame; a japan kaleidoscope; a wooden magic lantern; and an ivory-mounted opera glass. Here was a life of philosophical pursuits in 340 lots, broken up over the course of a couple of afternoons. By the time the hammer came down on the final bid (for a cluster of imitation fruit housed in a glass dome), the worldly goods of Alexander Tilloch had been dispersed to every corner of the thriving capital, as was only fitting for a man whose life was devoted to the circulation of ideas amid the growth of the scientific marketplace. He was a skeptic, an innovator, and a far-reaching publisher, and his readers and contributors were going to miss him, challenging and combative though he was. Luke Howard, above all, was aware of the debt he owed to the earliest patron of his work. Tilloch had played a crucial role in the life of the essay on clouds, and the impact that it had (and continues to have) was due as much to his editorial instincts as to the classifying insights of its author.

But by the time of Alexander Tilloch's death, the classification had taken on a life of its own. The great zoologist Thomas Huxley once commented that it was the customary fate of new scientific truths to begin as heresies and to end as superstitions, but this was not the case with Howard's clouds. The seven new names, with their poetic explanations, had, on the whole, been borne on waves of acclaim and acceptance from the outset. Yet as has already been seen, a number of reservations were voiced at the time of the essay's publication, most of which stemmed from simple disbelief that there could be only seven types of cloud—the atmosphere seemed so written over with an endless variety of types. This disbelief persisted even as the classification flourished, generating a series of suggested and disputed amendments to the scheme. The pace of these amendments quickened over the course of the nineteenth century as meteorologists began to organize themselves on an international footing. Consequently, by the end of the century, the shape of the classification as it is used today had largely been remodeled and reformed.

The first true modification to be added to Howard's list was "Stratocumulus," a term suggested by the German meteorologist Ludwig Kaemtz in 1840. Kaemtz, a professor of physics at the University of Halle, had sought to distinguish those rolling masses of forlorn gray cloud from what Howard had termed cumulostratus: "the Cirro-stratus blended with the Cumulus." Kaemtz's inversion of the terms removed the cloud form from the convective cumulus family and placed it in the category of stratus, reassigning it a more suitable position within the family of low-pressure clouds. Subsequently, and by general agreement, Howard's original term was dropped from the register. Stratocumulus, newly defined as "a layer of cloud, not flat enough to be called pure stratus, but rising into lumps too irregular and not sufficiently rocky to be called true cumulus," was the first major revision of Howard's earlier arrangement, although one that worked entirely within Howard's terminology, as well as with his idea of compound modification.[9]

Not long after, in 1855, Émilien Renou, director of the French observatories at Parc Saint-Maur and Montsouris, made two further additions to the classification, in the form of altocumulus and altostratus (*alto*, from the Latin for "elevated" or "high"). Both these new cloud formations, as Renou pointed out, were medium-level clouds, with their altitude emphasized, even in their name, as having the most shaping influence over their form.

This crucial observation gave renewed strength to the case for adopting altitude as the principal means for grading every family of cloud. Jean-Baptiste Lamarck had first proposed this idea in his meteorological almanac for 1802, but along with the rest of those ill-fated volumes, it had had little or no impact at the time. Half a century later, however, the idea came to life once more. Renou's recommendation was taken up by observatories all over Europe, where the emphasis on altitude as the defining quality of clouds grew steadily among their members. It was officially adopted for international use at the Paris Conference of 1896, the International Year of Clouds.[10]

Luke Howard's terms, meanwhile, continued to be added to and reordered. Some newly introduced categories, such as those devised by Kaemtz and Renou, were distinct new cloud genera in themselves: they were examples, that is, of what Howard would have called compound modifications. Other new categories were really secondary amendments, like those that had been devised by Forster in the 1810s, or the "broken" varieties, such as cumulus fractus, that were added in the 1860s by Andrès Poey in Havana. Howard, who continued to make his own amendments to the scope of his original definitions, would have expected nothing less than to see his classification refined and reorganized by his scientific contemporaries. And he would have been delighted to observe how the meaning and the impact of his original set of terms were preserved in these new acts of naming.

Except that by then Luke Howard was old and frail and preparing himself for death. His wife, Mariabella, had died in 1852 at the age of 83, following a swift and painful decline, and Luke was living with his eldest son, Robert, in a comfortable house in Tottenham. His memory had been failing over the previous twelve months and by the end, as Robert Howard recalled in his funeral oration, among the elements of his life that had been consigned to forgetfulness were the names of the clouds he had so lovingly bestowed some sixty years before:

> Those who lived with him will not soon forget his interest in the appearance of the sky. Whether at morning, noon, or night, he would go out to look around on the heavens, and notice the changes going on. His intelligent remarks and pictorial descriptions gave a charm to the scene never before realised by some. A beautiful sunset was a real and intense delight to him; he would stand at the window, change his position, go out of doors and watch it to the last lingering ray. It was a gratification to him to find a sympathizing admirer; long after he

> ceased, from failing memory, to name the 'cirrus', or 'cumulus', he would derive a mental feast from the gaze, and seem to recognise old friends in their outlines.[11]

How could he have lived without the clouds?

Luke Howard died at eleven o'clock in the evening on March 21, 1864, to the sound of his son's voice reading from the Book of Genesis: "and God said, This is the token of the covenant which I make between me and you and every living creature that is with you, for perpetual generations; I do set my bow in the cloud, and it shall be for a token of a covenant between me and the earth. And it shall come to pass, when I bring a cloud over the earth, that the bow shall be seen in the cloud: and I will remember my covenant."

The funeral service was held on March 26 at Winchmore Hill Burial Ground, an acre of tranquillity set high on the rim of the metropolis of London. He was placed next to his parents and his beloved Mariabella, his wife and partner of fifty-six years, and near the graves of his young half brothers, Robert and Joseph, who had died in the typhus outbreaks of the 1790s. The family was slowly reuniting. The service was well attended by his relatives and friends, who gathered in silence at the graveside. One of the grandchildren recalled soon after that "the time spent at the grave was a quiet and solemn occasion, leaving the impression on the minds of those present, that a good man had been gathered to his rest, after a longer pilgrimage than is usually alloted to man."[12]

The obituaries were united in recording not only the passing of an individual soul, of a grand old man of 91, but the closing of a well-respected age of science. "Never, probably, was Science wooed more entirely for her own sake," observed one; "never was there a more thorough 'labour of love' than that which he bestowed."[13] And the observation was true, it was a labor of love, as Luke Howard's grandson Thomas Hodgkin confirmed long after his grandfather's death: "I most like to think of him standing on the verandah and watching the dear clouds, the study of which had formed the delight of his life, and which, as I have said before, seemed to be especially beautiful for him."[14]

All who knew him were certain that he rested as and where he had always lived: in peace amid the gently rising clouds.

But by the middle of the nineteenth century the terms of Howard's original cloud classification had grown far beyond the limitations he had given them. More and more variations of usage were beginning to appear, particularly at overseas observatories, where local terms subtly introduced themselves alongside Howard's Latin. The French, for example, referred to cirro- and altocumulus as *ciel pommelé* or *ciel moutonné* (skies that were "dappled" or "fleecy"), while the Spanish described them as *cielo empedrado* (skies that were "cobbled" like a road). The same cloud structures in Britain, meanwhile, had long been known as a mackerel sky, from their resemblance to the scales of a fish. These local variants were harmless in themselves, of course, but because of reemerging confusions within the Latin terminology, they had begun to reappear in meteorological bulletins: pre-Howardian cloud designations were making a return to meteorology after an absence of more than half a century.

The whole system of cloud nomenclature was now in danger of fragmentation, especially since Howard's own terms had not escaped the spreading band of confusion. The same terms, although widely used, were not always being made to define the same cloud structures; and no classification of any kind could hope to withstand such usage. The terms cirrostratus and altostratus, for instance, were being used interchangeably, while cirrocumulus was known at a number of European observatories by the name of stratocumulus, the term that had been introduced by Ludwig Kaemtz in 1840 as a replacement for the disputed cumulostratus. It seems as if Kaemtz's intervention had been widely misunderstood, resulting in an entire group of cumuliform clouds being misrepresented as stratiform. Confusion at the observatories reigned, tempers grew heated, and the reapplication of demotic translations only aggravated the situation.

There were other questions, too, that needed definitive answers. Was fog a kind of stratus, as Howard had been keen to emphasize in his 1817 version of the classification, or was it not in fact a proper cloud at all? Could nimbus, the rain cloud, still be characterized as a cloud form in its own right? Did rain fall freely from an individual cloud, or did it need a combination of layers to begin? As long as these and other questions remained unanswered, as long as the certainty of knowing a cloud by a single, unambiguous, international name remained in doubt, the work begun by Luke Howard in 1802 would never be satisfactorily resolved. And now that he and most of his supporters were dead, what would become of his classification? Was it all about to come apart at the seams?

It was at this point that two men stepped forward to take charge of the whole situation: Professor H. Hildebrand Hildebrandsson of the University Observatory of Uppsala, Sweden, and the Hon. Ralph Abercromby of Belgrave Square and the Royal Meteorological Society. Both had attended the first International Meteorological Congress in 1873, and both were organizational geniuses fired with a passion for clouds. They were determined, above all, to see nephological differences reconciled around the world for the sake of harmony and progress in the science. The pair resolved that, between them, they would umpire the entire range of claimed classifications, whether new or established, and then pronounce, once and for all, on behalf of the world meteorological community, the final word on the permanent language of clouds. It was a bold undertaking, but given the circumstances that had begun to develop long before Luke Howard's death, it was the only reasonable step they could take. They would, they declared, establish a global bureaucracy of naming.

They certainly had the right qualifications. Hildebrandsson, who bore a striking resemblance to the elderly Karl Marx, had already published a short cloud classification for the use of his students at the University of Uppsala. In it, he praised Luke Howard for his "precise observations" before going on to point out that a number of confusions had arisen within the terminology.[15] These, he felt, were in need of urgent attention, lest the clouds slip free from the leash of Howard's words. Something needed to be done. Abercromby had long thought the same thing himself, and when he came across the passage in Hildebrandsson's book, he was delighted to have found a sympathetic ally. He traveled to Sweden to discuss the matter with Hildebrandsson, and published the results of their lengthy conversations in the *Quarterly Journal* of the recently renamed Royal Meteorological Society in 1887. The pair had gotten on famously, for Abercromby had also made a name for himself in meteorological circles, but in a far more spectacular way than the Swede. As a wealthy, aristocratic Englishman, complete with a fulsomely waxed mustache, Abercromby had employed his ample time and financial resources in voyaging twice around the globe by steamer, train, and carriage, making sure that the clouds did indeed look the same the world over, from New Zealand to Newcastle, via America, Borneo, and the high Himalayas. Abercromby's voyages had taken years to complete, and there was more than a touch of Phileas Fogg about him, an impression cheerfully bolstered by the written account of his travels, which he published in 1888: *Seas and Skies in Many Latitudes, or, Wanderings in Search of Weather.*

But though his means were great, his ends, as he saw them, were modest enough in scope, according exactly with the aims of the older Hildebrandsson: "My primary idea was that the name of a cloud is of far less importance than that the same name should be applied to the same cloud by all observers," he wrote.[16] The pair was in total agreement on the matter, and at their meeting in Sweden in 1887, Hildebrandsson and Abercromby drew up a list of ten provisional cloud types, "all compounded of Howard's four fundamental types—*Cirrus, Stratus, Cumulus, Nimbus,*—" that would "fully meet the requirements of practical meteorology." These were then printed and offered up "for the consideration of meteorologists" everywhere.[17]

The ten terms—Cirrus, Cirro-stratus, Cirro-cumulus, Strato-cirrus, Cumulo-cirrus, Strato-cumulus, Cumulus, Cumulo-nimbus, Nimbus, and Stratus—have gone on to form the basis of every subsequent arrangement of clouds, and in some respects this end-of-the-century declaration proved just as influential as Howard's essay had on its first publication at the century's start. Because of the efforts of Hildebrandsson and Abercromby, a pair of international men of weather, clouds had been returned to the forefront of world meteorological thought.

This reorganization of the terminology of clouds was to bear immediate fruit. The publication of a multilingual cloud directory was proposed by an international team in Hamburg as a demonstration piece for the new classification.[18] The project was led by Hildebrandsson from his office at the Swedish Observatory, and when the book appeared in 1890 its significance as "the first satisfactory attempt to attain uniformity in the classification and nomenclature of clouds" was immediately recognized.[19] The introductory text, based closely on the 1887 paper, was in four languages—German, French, English, and Swedish—and was followed up by ten chromolithographs and two pages of tipped-in photographs of a strikingly professional quality, most of which had been taken by Abercromby himself during his extensive travels overseas.

Photography was an impressive new technical advance that had only recently come into prominent scientific usage, but it was soon to fill a major evidentiary role in the newly drawn domain of nephology. Abercromby had devoted more than ten years to perfecting the difficult art of photographing clouds, an activity that he held to be the only possible route toward securing international agreement on the subject. "No international accordance of cloud names can be obtained until typical photographs can be circulated at a moderate price," he claimed, particularly if those photographs were his own.[20] Abercromby's pictures were genuinely admired, however, and his evening slide shows at the Westminster home of the Royal Meteorological Society were well-attended events. Their impact was profound, one en-

Photograph of cumulus by Ralph Abercromby

thusiast describing the photographs afterward as "the most beautiful cloud pictures he had ever seen."[21] No one in London in the 1880s ever said that about Constable's.

Following the publication of the Hamburg atlas, the next International Meteorological Conference decided to devote itself to the subject of clouds. Bureaucratic to a fault, it elected a Cloud Committee, whose members were to chair the discussions. The talks, held in Munich throughout September 1891, were long and wide-ranging, but the chief resolution that was finally reached was the acceptance of the new ten-point cloud classification as proposed by Hildebrandsson and Abercromby, and as demonstrated in the Hamburg atlas. Its merits were obvious to all who examined it. It updated Howard's original classification while keeping strictly to his long-familiar nomenclature.

Fired with enthusiasm, the members of the Cloud Committee arranged to meet again in Uppsala in 1894 to curate an exhibition of cloud pictures. These were to be collected from around the world through advertisements and appeals in the journals; the ad that appeared in the *American Meteorological Journal* of March 1892 was a typical example:

> A REQUEST FOR CLOUD PICTURES—The International Meteorological Conference, which met at Munich last year, decided to publish a col-

> ored Cloud Atlas representing the typical cloud forms according to the nomenclature of Hildebrandsson and Abercromby. The committee appointed to carry out the project, request the loan of colored drawings or paintings made from nature, in order that the most suitable may be reproduced in the Atlas. Such cloud studies may be sent for inspection to the American member of the commission, A. Lawrence Rotch, Blue Hill Observatory, Readville, Mass., and will be returned in good condition to the lenders.[22]

Thousands of pictures were sent in to the show, with well over three hundred finally being accepted. The Uppsala loan exhibition was a great success and traveled subsequently to Britain and North America. Not only did it provide a further opportunity for the members of the Cloud Committee to engage in some congenial tourism, but it operated as a selection exercise to aid the final decision on the illustrations for a new publication: the first *International Cloud Atlas*, which was to be published in 1896.

The new atlas, of course, had its own committee (an elected subgroup of the Cloud Committee), formed for the purpose of speeding up events. The book was intended to be an improvement on the earlier Hamburg edition, especially in the choice of illustrations, but it needed to be published in time to coincide with the

Meteorological Conference of 1896. For as a means to publicize the event, and to mark its significance, 1896 had been declared the International Year of Clouds. Meteorologists everywhere were patiently waiting for their copies of the commemorative guidebook.

They were not kept waiting long, and soon after it appeared in early summer, the slim trilingual cloud atlas, published in blue boards by the Gauthier-Villars press of Paris, had made its way to every corner of the scientific world.

To compare the classification of 1802 with the version published nearly a century later, in 1896, is to be reminded of just how significant were the changes made in the course of that time. Clouds were now classified first by height, into "Upper Clouds," "Intermediate Clouds," "Lower Clouds," "Clouds of Diurnal Ascending Currents," and "High Fogs"; then, within each height category, there were two further divisions: "*a*. Separate or globular masses (most frequently seen in dry weather)" and "*b*. Forms which are widely extended, or completely cover the sky (in wet weather)." The ten cloud genera, according to the *Atlas*, were now to be ranked as follows:

A. **Upper Clouds**, average altitude 9000^{m}.

a. 1. *Cirrus.*

b. 2. *Cirro-stratus.*

B. Intermediate Clouds, between 3000^{m} and 7000^{m}.

a. 3. *Cirro-cumulus.*

b. 4. *Alto-cumulus.*

b. 5. *Alto-stratus.*

C. Lower Clouds, 2000^{m}.

a. 6. *Strato-cumulus.*

b. 7. *Nimbus.*

D. Clouds of Diurnal Ascending Currents.

8. *Cumulus*; apex, 1800^{m}; base, 1400^{m}.

9. *Cumulo-nimbus*; apex, 3000^{m} to 8000^{m}; base, 1400^{m}.

E. High Fogs, under 1000^{m}.

10. *Stratus.*[23]

This reorganized classification, with its generous array of accompanying pictures, ensured that every visible cloud at any given altitude could now be named with confidence by any observer in the world.

This fact was not lost on the meteorological community. "We strongly recommend our readers to procure a copy," advised the London-based *Meteorological Magazine* in July 1896; "the atlas is very lovely and very cheap." The review went on to note that while the *Atlas* had updated the contents of Luke Howard's original essay, it had made sure to preserve the language and the ideas that had stayed familiar to their users over the better part of the century that had elapsed since their inception: "Our countrymen may well be content to see how largely the International System of 1896 is based upon the work of Howard."[24]

And it was true—his countrymen were content—for now, at last, it was official: Luke Howard had named the clouds, for all countries, all peoples, and all time.

Epilogue

Afterlife

But the simple work of classification of the clouds, depending chiefly on a quick eye for form and colour, and on the possession of the philosophic habit of mind, still remains; and all over the globe, wherever scientific observers are to be found, the clouds are still known by the names whereby he named them.

The Friend, 1864[1]

"My grandfather was a sensitive delicate man, with a good deal of the oddity of genius, and its waywardness," recalled Luke Howard's granddaughter Mariabella Fry, toward the end of her own life in the 1920s. Her depiction of the gentle, elderly figure was vivid, affectionate, and true to the life of the man she remembered: "He was a very absent man, and seemed always to be thinking of something very far away, so that we seldom ventured into conversation with him, and if we asked a question we were often met with the vague 'my dear' which showed that he had not been attending. He was often contemplating the weather and would stand for a long time at the window gazing at the sky with his dreamy placid look, occasionally drawing our attention to some grand cloud, and explaining its form. He used to say 'people think I am *weather-wise*, but I tell them I am very often *otherwise*.' "[2]

The reference to Benjamin Franklin was a telling one: "Some are weather-wise; some are otherwise" was one of the hundreds of homespun sayings dispensed throughout Franklin's *Poor Richard's Almanac*, the set of handbooks of mercantile philosophy that also gave us "Time is money," "Little strokes fell great oaks," and "Love your neighbor, but don't pull down your hedge." Penny-pinching and pious, Franklin's almanac was the barometer of Protestant anxiety, and Luke Howard, like so many of his kind, applied himself with diligence to its lessons.

Though inclined to daydreaming and to staring out the window, Howard had "a deep and anxiety-making need to be useful," as Nicholas Webb has phrased it, and wrote letters to his friends in which "he castigated himself for leaving his days 'so utterly void of end or aim or usefulness' ":

> My desire is that whenever and so long as I may be fit for exertion I may be directed to use it for better purposes than my own gratification or advantage in point of outward wealth. Yet *how* to be more useful to others I do not yet clearly see; so I keep still in my corner.[3]

This was a self-assessment written not at the quiet end of his retirement in the 1850s or 1860s, but in 1811, at the height of his engagement in business, science, and family life—about as far from keeping still in his corner as it was possible to go. His manufacturing laboratory was at full production, he was submitting monthly weather bulletins to a variety of journals, he was preparing the preliminary drafts of *The Climate of London*, and he was agitating on behalf of the victims of the Napoleonic campaigns. He was also much involved in the lives of his seven children, teaching French to the older ones in the mornings before work and taking them on regular weekend study trips to Stonehenge or the British Museum. Here, then, was a family man with a busy, well-rounded life, a man fully embarked on the Protestant journey into substance. But his anxieties over his value as a citizen and as a man haunted him nonetheless, for he was, after all, his father's son and had a deep-rooted terror of idleness.

Luke Howard was a singular product of his age and social background: a self-employed small-businessman with an unusual passion for clouds, whose sole qualifications were his schoolboy Latin and a membership of the Plough Court academy. Yet his early work on the modifications of clouds went on to change the face of meteorology forever, earning the admiration of his leading contemporaries in literature, art, and science. By naming the clouds, by giving language and a greater visibility to things that had hitherto been nameless and unknowable, he completely transformed the relationship between the world and its overarching sky. His mutable science, the science of the shape-shifting clouds, spoke clearly and comfortingly to an era bewildered by the pace of social and industrial change. "The ocean of air in which we live and move, with its continents and islands of cloud," he once wrote, "can never to the conscious mind be an object of unfeeling contemplation."

Thirty-two years after his death, his achievement was crystallized in the pages of the *International Cloud Atlas*, at the time the world's most authoritative meteorological publication: "No one can examine this Atlas or study the subject without seeing that the foundation was laid by Luke Howard," as the *Quarterly Journal* of the Royal Meteorological Society put it in 1896; "if ever a man deserved recognition in connection with this subject, it is this famous scientist and noble philanthropist."[4] And that was only the beginning. Since its first appearance in 1896, the *International Cloud Atlas* has had seven further English-language editions, the most recent of which appeared in two volumes in 1995. The latest coeditions, in French and Spanish, appeared in 1975 and 1993, respectively. Still based on Howard's foundational work, the *Atlas* remains the definitive guide to the international classification and nomenclature of clouds.

The cloud terms that are currently in use, with their international abbreviations and the dates of their agreed definition (or redefinition) are now arranged as follows:

High Clouds (their bases above 6 km):

1) *Cirrus, Ci* (Howard, 1803)

2) *Cirrocumulus, Cc* (Howard, 1803; Renou, 1855)

3) *Cirrostratus, Cs* (Howard, 1803; Renou, 1855)

Middle Clouds (their bases between 2 and 6 km):

4) *Altocumulus, Ac* (Renou, 1870)

5) *Altostratus, As* (Renou, 1877)

6) *Nimbostratus, Ns* (International Commission for the Study of Clouds, 1930)

Low Clouds (their bases below 2 km):

7) *Stratocumulus, Sc* (Kaemtz, 1841)

8) *Stratus, St* (Howard, 1803; Hildebrandsson and Abercromby, 1887)

9) *Cumulus, Cu* (Howard, 1803)

The largest cloud-type of all regularly stretches right through the three altitude bands:

10) *Cumulonimbus, Cb* (Weilbach, 1880)

As can be seen, the current classification is less complex than it was in 1896, with a simpler, tripartite, altitude structure based partly on Lamarck's earlier conception. Only the names of the ten principal cloud forms, the ten genera listed above, are regularly used in meteorological bulletins. The appendix on p. 355 gives the full range of currently recognized species and varieties, to which further additions will no doubt be made in the future. And, sad to report, cumulonimbus is no longer cloud nine.

Cloud studies continue to flourish around the world, especially now that research has confirmed that clouds play a far greater role in regulating climate and climatic change than was previously suspected. The findings of groups such as C^4 in California (the Center for Clouds, Chemistry, and Climate) demonstrate the importance of clouds in screening sunlight from the earth and in accelerating the radiation of heat away from it. Clouds, in other words, act as a vast and efficient global cooling system, with different cloud types appearing to do their work in different ways. Bands of stratus cloud tend to reflect light back into the stratosphere, for example, while the ice crystals within cirriform clouds tend to absorb radiated heat from below. The extent of this thermostatic cooperation between the earth and its circulating atmosphere is only just beginning to be understood, but clouds are clearly at the center of it all.

"Clouds always tell a true story," as Ralph Abercromby said in 1887, "but one which is difficult to read."[5] And clouds are going to go on telling their stories, endlessly and forever. Yet because so much about them remains to be learned, because so much about them is unknown and unknowable, they will never lose their mystery or their majesty. Luke Howard's science has only helped us to read them and to enjoy their exhibitions all the more: it allows for daydreaming, for watching clouds pass, whether lying on our backs on a sunny summer's day, or driving through the mountains with their mist-enveloped peaks. None of our delight and none of our reverence has been lost. Clouds surround us, clouds give us comfort, and like Howard, or Abercromby, or like Baudelaire's extraordinary stranger, we can all fall a little in love with them.

So let's look at the clouds . . . the drifting clouds . . . there . . . over there . . . the marvelous clouds.

Appendix

Cloud Species and Varieties

Species

Within the ten cloud genera, as listed on p. 351, there are fourteen cloud species, describing the shape and structure of each type of cloud. Each specific term could in theory be applied to any one of the ten genera, but in practice they are usually applied to only one or two. The terms date mostly from the mid–twentieth century onward:

Typical Genera	Species	Abbr.	Description
Cirrus	*uncinus*	*unc*	*comma- or hook-shaped*
Cirrus	*spissatus*	*spi*	*dense and gray toward the sun*
Ci, Cs	*fibratus*	*fib*	*nearly straight, no hooks*
Sc, Ac, Cc	*lenticularis*	*len*	*wave cloud, almond- or lens-shaped*
Sc, Ac, Cc, Ci	*castellatus*	*cas*	*turrets with a common base*
Ac, Cc, Ci	*floccus*	*flo*	*small tufts of cloud, ragged lower part*
Cumulonimbus	*capillatus*	*cap*	*distinct icy region with anvils or plumes*
Cumulonimbus	*calvus*	*cal*	*tops look smooth or bald*
Sc, Ac, Cc	*stratiformis*	*str*	*extensive horizontal sheet or layer*
St, Cs	*nebulosus*	*neb*	*thin veil or layer*
Cumulus; stratus	*fractus*	*fra*	*ragged shreds of cloud*
Cumulus	*congestus*	*con*	*growing, with cauliflower-shaped tops*
Cumulus	*mediocris*	*med*	*moderate depth, tops with small bulges*
Cumulus	*humilis*	*hum*	*flattened*

Varieties

The cloud variety defines either the transparency of the genera or species, or the particular arrangement of its elements:

Typical Genera	Variety	Abbr.	Description
Cirrus	*intortus*	*in*	*irregularly curved or tangled*
Cirrus	*vertebratus*	*ve*	*like ribs or fish bones*
Sc, Ac, As, Cc, Cs	*undulatus*	*un*	*with parallel undulations*
Cu, Sc, Ac, As, Ci	*radiatus*	*ra*	*parallel bands seeming to converge*
Ac, Cc	*lacunosus*	*la*	*with holes, reticulated like a net*
Sc, Ac, As, Ci, Cs	*duplicatus*	*du*	*more than one layer*
St, Sc, Ac, As	*translucidus*	*tr*	*translucent, showing the sun or moon*
Sc, Ac	*perlucidus*	*pe*	*allows sun- or moonlight to be seen*
St, Sc, Ac, As	*opacus*	*op*	*completely masks sun or moon*

Accessory Clouds

Certain forms of cloud are not types in themselves, but appearances occurring only in association with one or two of the genera. They can often suggest physical processes occurring within a cloud. The nine accessory clouds are:

pannus	*shreds of cloud, occurring with Cu, Cb, As, Ns*
pileus	*cap cloud, occurring with cumulus and cumulonimbus*
velum	*veil, occurring with cumulus and cumulonimbus*
arcus	*arch cloud*
incus	*anvil cloud*
mamma	*pouches hanging down from upper cloud*
praecipitatio	*precipitation reaching the surface*
tuba	*funnel clouds*
virga	*fallstreaks; precipitation not reaching the surface*

Meteorologists employ all the above terms to define and describe clouds, but weather bulletins usually restrict themselves to the names of the ten genera. And in general conversation about the state of the sky, we tend to use only the three initial family names as defined by Luke Howard in 1802: cirrus, stratus, and cumulus.

Notes

Prologue: The Useless Pursuit of Shadows

1. Charles Baudelaire, *The Parisian Prowler: Le Spleen de Paris, Petits Poèmes en prose,* trans. Edward K. Kaplan (Athens, Ga.: University of Georgia Press, 1989), p. 1.
2. Based on Luke Howard, *On the Modifications of Clouds, &c.* (London: J. Taylor, 1804), p. 3.

1. The Theater of Science

1. Sarah Hoare, *Poems on Conchology and Botany* (London: Simpkin & Marshall, 1831), p. 61.
2. Richard D. Altick, *The Shows of London* (Cambridge, Mass.: Harvard University Press, 1978).
3. *Dictionary of Scientific Biography*, s.v. "Lavoisier, Antoine-Laurent."
4. David Knight, *Humphry Davy* (Oxford: Blackwells, 1992), p. 50.
5. *The Parachute; or, All the World Balloon Mad* (London: for the author, 1802), p. 2.
6. Kathleen Coburn, "Coleridge: A Bridge between Science and Poetry," in *Coleridge's Variety: Bicentenary Studies*, ed. John Beer (London: Macmillan, 1974), pp. 87–90.
7. John Ayrton Paris, *The Life of Sir Humphry Davy* (London: Colburn & Bentley, 1831), p. 90.
8. See John Emsley, *The Shocking History of Phosphorus: A Biography of the Devil's Element* (London: Macmillan, 2000).
9. Samuel Johnson, *The Idler* (London: J. Newbery, 1761), 2: 59.
10. Arden Reed, *Romantic Weather: The Climates of Coleridge and Baudelaire* (Hanover and London: Brown University Press, 1983), p. 6.
11. Huntington MS, MO 3055 (with thanks to Elizabeth Eger for the reference).
12. *The Works of Jane Austen*, ed. R. W. Chapman (Oxford: Oxford University Press, 1954), 6: 403.
13. Hoare, p. 7.
14. Richard Holmes, *Coleridge: Early Visions* (London: Hodder & Stoughton, 1989), p. 312.
15. Coburn, p. 85.

2. A Brief History of Clouds

1. Aristophanes, "The Clouds," in *Lysistrata and Other Plays*, trans. Alan H. Sommerstein (Harmondsworth: Penguin, 1973), p. 125.
2. Gustav Hellmann, "The Dawn of Meteorology," *Quarterly Journal of the Royal Meteorological Society* 34 (1908): 223.
3. "The Great Summons," in *Chinese Poems*, ed. Arthur Waley (London: Allen and Unwin, 1946), p. 37.
4. Colin A. Roman, *The Cambridge Illustrated History of the World's Science* (Cambridge: Cambridge University Press, 1983), pp. 125–86.
5. Sir Napier Shaw, *Manual of Meteorology* (Cambridge: Cambridge University Press, 1926), 1: 5–6; see also J. A. Kington, "A Historical Review of Cloud Study," *Weather* 23 (1968): 349–56.
6. Kevin Crossley-Holland, *The Norse Myths: Gods of the Vikings* (London: Deutsch, 1980), p. 144.
7. See H. Howard Frisinger, *The History of Meteorology: to 1800* (New York: Science History Publications, 1977), p. 5; Jonathan Barnes, *Early Greek Philosophy* (Harmondsworth: Penguin, 1987), pp. 71–6.
8. Hellmann, "Dawn of Meteorology," p. 224.
9. The best concise account of this is in David C. Lindberg, *The Beginnings of Western Science: The European Scientific Tradition in Philosophical, Religious, and Institutional Context, 600 B.C. to A.D. 1450* (Chicago and London: University of Chicago Press, 1992), pp. 46–68.

10. Aristotle, *Meteorologica*, trans. H.D.P. Lee (London: Heinemann, 1952), p. 10.
11. Ibid., p. 17.
12. Frisinger, *History of Meteorology*, pp. 15–17.
13. Seneca, *Naturales Quaestiones*, trans. T. H. Corcoran (London: Heinemann, 1971), 2: 273.
14. Pliny, *Natural History*, trans. H. Rackham (London: Heinemann, 1937), 1: 247.
15. Lucretius, *On the Nature of the Universe*, trans. R. E. Latham, rev. John Godwin (Harmondsworth: Penguin, 1994), p. 178.
16. Cited in Jack Rochford Vrooman, *René Descartes: A Biography* (New York: Putnam, 1970), pp. 123–4.
17. Anthony Le Grand, *An Entire Body of Philosophy, According to the Principles of the Famous Renate Des Cartes, in Three Books* (London: Samuel Roycroft, 1694), p. 213.
18. Le Grand, p. 215. See also Claire L. Parkinson, *Breakthroughs: A Chronology of Great Achievements in Science and Mathematics 1200–1930* (London: Mansell, 1985), p. 79.
19. Oliver Goldsmith, *A History of the Earth, and Animated Nature*, new ed. (London: Wingrove and Collingwood et al., 1816), 1: 315–20.
20. Charles Taylor, *Surveys of Nature, Historical, Moral, and Entertaining, exhibiting the Principles of Natural Science* (London: C. Taylor, 1787), 1: 192.
21. Goldsmith, 1: 312.

3. The Cloud Messenger

1. Mary Russell Mitford, "Song," in *Poems*, 2d ed. (London: A. J. Valpy, 1811), p. 107.
2. Luke Howard, *On the Modifications of Clouds, &c.* (London: J. Taylor, 1804), p. 4.
3. Daniel Defoe, *The Storm: or, A Collection of the most remarkable Casualties and Disasters Which happen'd in the Late Dreadful Tempest, both by Sea and Land* (London: G. Sawbridge, 1704), p. i.

4. Scenes from Childhood

1. Bernard Howard, "A Luke Howard Miscellany: Compiled by his Great Grandson" (London: unpublished typescript, 1959), p. 86 (Friends' House Library, Euston Road, London).
2. LHM, p. 59.
3. LHM, pp. 60–1.
4. London Metropolitan Archive: Acc. 1017/1381.
5. LMA: Acc. 1017/1372–1388.
6. LHM, pp. 62–3.
7. *Short Memorials of the late Luke and Mariabella Howard, of Ackworth Villa, Yorkshire*, by an aged relative (Tottenham: privately printed, 1864), p. 3.
8. Obituary from *The Friend: A Religious, Literary and Miscellaneous Journal*, new series 4 (London: 1864): 99–102.

9. Gilbert White, *The Natural History and Antiquities of Selborne*, ed. Paul Foster (Oxford: Oxford University Press, 1993), pp. 247–8.

10. This section has been compiled from contemporary newspaper reports from April to October 1783: *Bath Chronicle, Bonner & Middleton's Bristol Journal, Cambridge Chronicle & Journal, Canterbury Journal, Jopson's Coventry Mercury, Caledonian Mercury* (Edinburgh), *Gloucester Journal, Gentleman's Magazine* (London), *London Gazette, Norfolk Chronicle or Norwich Gazette, Jackson's Oxford Journal*, and *York Courant.* See also Richard Mabey, *Gilbert White: A Biography of the Author of The Natural History of Selborne* (London: Century, 1986), pp. 190–9; John Grattan and Mark Brayshay, "An Amazing and Portentous Summer: Environmental and Social Responses in Britain to the 1783 Eruption of an Iceland Volcano," *The Geographical Journal*, 161, no. 2 (1995): 119–34; and Richard B. Stothers, "The Great Dry Fog of 1783," *Climatic Change* 32 (1996): 79–89.

11. *The Letters and Prose Writings of William Cowper*, ed. James King and Charles Ryskamp (Oxford: Clarendon Press, 1979–86), 2: 144.

12. *The Letters of Horace Walpole, fourth Earl of Orford, Chronologically arranged and edited with notes and indices by Mrs Paget Toynbee* (Oxford: Clarendon Press, 1903–25), 13: 12.

13. *Gentleman's Magazine*, vol. 53, no. 2 (1783): 621.

14. *Jackson's Oxford Journal*, 12 July 1783, p. 1.

15. *Lady's Magazine; or, Entertaining Companion for the Fair Sex, appropriated solely to their Use and Amusement*, vol. 17 (1786), p. 680.

16. Hayman Rooke, *A Meteorological Register, kept at Mansfield Woodhouse, in Nottinghamshire; from the commencement*

of the year 1785, to the end of the year 1794. To which are subjoined, The most probable Indications of Weather, deducible from the Changes in the Barometer (Nottingham: Samuel Tupman, 1795); updated annually until 1805.

17. Hayman Rooke, *A Continuation of the Annual Meteorological Register, kept at Mansfield Woodhouse, from the year 1800 to the end of the year 1801* (Nottingham: Samuel Tupman, 1802), p. 22.
18. John and Mary Gribbin, *Watching the Weather* (London: Constable, 1996), p. 1.
19. H. H. Lamb, "Volcanic dust in the atmosphere; with a chronology and assessment of its meteorological significance," *Philosophical Transactions of the Royal Society* A266 (1970): 425–533.
20. Although the VEI was developed in 1982, it is not yet the globally accepted standard. Volcanology, unlike seismology, continues to rely partly on subjective measurements. See Tom Simkin and Lee Siebert, *Volcanoes of the World*, 2d ed. (Tucson, Ariz.: Geoscience, 1994), pp. 23–5.
21. Quoted by H. H. Lamb, p. 509.
22. Ibid.
23. Stothers, p. 79.
24. See "Fatalities & Evacuations," in Simkin and Siebert, pp. 165–75; Stothers, p. 85.
25. John Pointer, *A Rational Account of the Weather; Shewing the Signs of its several Changes and Alterations, together with the Philosophical Reasons of them* (London: Aaron Ward, 1723), p. iii.
26. Pointer, p. iv.
27. LMA: Acc. 1017/1377.
28. LMA: Acc. 1017/1376.
29. LHM, p. 62.

30. "Memoir of Luke Howard," in *Luke Howard (1772–1864) His correspondence with Goethe and his Continental Journey of 1816*, ed. D.F.S. Scott (York: William Sessions, 1976), p. 2.
31. LMA: Acc. 1017/1385.

5. The Askesian Society

1. "Luke Howard's Autobiography; with his own additions and corrections down to an unascertained date. Probably circ. 1840" (Friends' House Library, MS box 5.2), p. 5.
2. Bernard Howard, "A Luke Howard Miscellany: Compiled by his Great Grandson" (London: unpublished typescript, 1959), p. 61.
3. John Emsley, *The Shocking History of Phosphorus: A Biography of the Devil's Element* (London: Macmillan, 2000).
4. LHM, p. 94.
5. *Short Memorials of the late Luke and Mariabella Howard, of Ackworth Villa, Yorkshire*, by an aged relative (Tottenham: privately printed, 1864), p. 6.
6. LHM, p. 94.
7. *The Collected Works of Samuel Taylor Coleridge*, ed. H. J. Jackson and J. R. de J. Jackson (London: Routledge, 1971–), 11: 335–6.
8. Desmond King-Hele, *Erasmus Darwin and the Romantic Poets* (London: Macmillan, 1986), p. 67.
9. John Thelwall, *The Peripatetic; or, Sketches of the Heart, of Nature and Society* (London: for the author, 1793) 2: 105–6.

10. An unpublished document in the London Metropolitan Archives, Acc. 1017/1491, written by William Haseldyne Pepys, contains an outline history of the Askesian Society. There is also a brief account in Luke Howard, *Papers on Meteorology, Relating Especially to The Climate of Britain, and to The Variations of the Barometer* (London: Taylor and Francis, 1854), pp. 73–6. See also Ian Inkster, "Science and Society in the Metropolis: A Preliminary Examination of the Social and Institutional Context of the Askesian Society of London, 1796–1807," *Annals of Science* 34 (1977): 1–32.
11. "Memoir of Luke Howard," in *Luke Howard (1772–1864) His correspondence with Goethe and his Continental Journey of 1816*, ed. D.F.S. Scott (York: William Sessions, 1976), p. 3.
12. *Life of William Allen, with Selections from his Correspondence* (London: Charles Gilpin, 1847), 1: 46–7.
13. Ibid., pp. 47–8.
14. David Knight, *Humphry Davy* (Oxford: Blackwells, 1992), pp. 30–1.
15. W.D.A. Smith, *Under the Influence: A History of Nitrous Oxide and Oxygen Anaesthesia* (London: Macmillan, 1982), p. 34.
16. "Memoir of Luke Howard," p. 3.
17. LHM, pp. 138–9.
18. L.T.C. Rolt, *The Aeronauts: A History of Ballooning 1783–1903* (Gloucester: Alan Sutton, 1985), p. 52.
19. Rolt, p. 54.
20. Richard Mabey, *Gilbert White: A Biography of the Author of The Natural History of Selborne* (London: Century, 1986), p. 196.
21. *Air-Balloon, or Blanchard's Triumphal Entry into the Etherial World: A Poem* (c. 1785), p. 16; Mary Alcock, *The Air Balloon; or, Flying Mortal, A Poem* (London: E. Macklew, 1784), p. 4.

22. Peter Brimblecombe, "Earliest atmospheric profile," *New Scientist* (1977): 364–5.

23. *The Balloon; or Aerostatic Spy: A Novel, Containing a Series of Adventures of an Aerial Traveller; Including a Variety of Histories and Characters in Real Life*, 2 vols. (London: W. Lane, 1786).

24. LHM, p. 144.

25. *The Parachute; or, All the World Balloon Mad* (London: for the author, 1802), p. 2.

26. Thomas Baldwin, *Airopaidia: Containing the Narrative of a Balloon Excursion from Chester, the eighth of September, 1785, taken from minutes made during the Voyage*, etc. (Chester: for the author, 1786) 2: 49.

27. *The Diary of John Evelyn*, ed. E. S. de Beer (Oxford: Clarendon Press, 1955) 2: 207–8.

28. See Jamie James, *The Music of the Spheres: Music, Science, and the Natural Order of the Universe* (New York: Grove Press, 1993), pp. 98–113; Leslie Orrey, *Opera: A Concise History* (London: Thames and Hudson, 1972), pp. 28–9.

6. Other Classifications

1. Jonathan Swift, *A Tale of a Tub and Other Works* (Oxford: World's Classics, 1986), p. 16.

2. John Kington, *The Weather of the 1780s over Europe* (Cambridge: Cambridge University Press, 1988), p. 4.

3. Thomas Sprat, *The History of the Royal-Society of London, For the Improving of Natural Knowledge* (London: J. Martyn, 1667), p. 179.

4. Luke Howard, *On the Modifications of Clouds, &c.* (London: J. Taylor, 1804), p. 3.

5. Sprat, p. 177.
6. Ibid., p. 174.
7. Ibid., p. 177.
8. Humphrey Jennings, *Pandaemonium 1660–1886: The coming of the machine as seen by contemporary observers*, ed. Mary-Lou Jennings and Charles Madge (London: Deutsch, 1985), p. 10.
9. J. A. Kington, "The Societas Meteorologica Palatina: An Eighteenth-Century Meteorological Society," *Weather* 29 (1974): 416–26.
10. Kington, *Weather of the 1780s*, p. 12.
11. L. J. Jordanova, *Lamarck* (Oxford University Press, 1984), pp. 11–19.
12. Pietro Corsi, *The Age of Lamarck: Evolutionary Theories in France 1790–1830* (Berkeley: University of California Press, 1988), p. 59.
13. Ibid., p. 60.
14. Ibid., p. 61.
15. Kh. Khrgian, *Meteorology: A Historical Survey*, 2d ed., trans. Ron Hardin (Jerusalem: Israel Program for Scientific Translations, 1970), p. 91.
16. Roger Clausse and Léopold Facy, *The Clouds*, trans. Joan Ferrante (New York: Grove Press, 1961), p. 21.
17. F. H. Ludlam, "History of Cloud Classifications," in Richard Scorer, *Clouds of the World: A Complete Colour Encyclopedia* (Newton Abbot: David & Charles, 1972), p. 17.
18. Gordon Brotherston, "The Republican Calendar: A Diagnostic of the French Revolution," in

1789: *Reading Writing Revolution*, ed. Francis Barker et al. (Colchester: University of Essex, 1982), pp. 1–11.

19. Mona Ozouf, *Festivals and the French Revolution*, trans. Alan Sheridan (Cambridge, Mass.: Harvard University Press, 1988), p. xiv.
20. Cited in E. G. Richards, *Mapping Time: The Calendar and Its History* (Oxford: Oxford University Press, 1998), p. 261.
21. Hayman Rooke, *A Continuation of the Annual Meteorological Register, kept at Mansfield Woodhouse, from the year 1802 to the end of the year 1803* (Nottingham: Samuel Tupman, 1804), p. 28.
22. *Nouveau Dictionnaire d'Histoire Naturelle* (Paris: 1818), 20: 451–77.

7. Publication

1. James Thomson, "The Castle of Indolence," in *Poetical Works*, ed. J. Logie Robertson (Oxford: Oxford University Press, 1908), p. 272.
2. G. J. Symons, "The History of English Meteorological Societies, 1823 to 1880," *Quarterly Journal of the Meteorological Society* 7 (1881): 73.
3. *The Letters of Robert Burns*, ed. G. Ross Roy (Oxford: Oxford University Press, 1985), 1: 395.
4. Luke Howard, *On the Modifications of Clouds, &c.* (London: J. Taylor, 1804), p. 21.
5. Bernard Howard, "A Luke Howard Miscellany: Compiled by his Great Grandson" (unpublished typescript, 1959), p. 138.

6. Luke Howard, *Seven Lectures on Meteorology* (Pontefract: James Lucas, 1837), pp. 84–5.
7. John A. Day and Frank H. Ludlam, "Luke Howard and his Clouds: A Contribution to the Early History of Cloud Physics," *Weather* 27 (1972): 449.
8. Luke Howard, *Modifications of Clouds*, p. 4.
9. L. J. Jordanova, *Lamarck*, p. 62.
10. Luke Howard, *Modifications of Clouds*, p. 4.
11. Ibid., pp. 5–6.
12. Ibid., pp. 6–11.
13. Luke Howard, *Seven Lectures*, p. 89.
14. Ibid.
15. Richard W. Burkhardt, Jr., *The Spirit of System: Lamarck and Evolutionary Biology* (Cambridge, Mass.: Harvard University Press, 1977), p. 17.
16. Luke Howard, *Modifications of Clouds*, p. 14.

8. Growing Influence

1. Charlotte Smith, *Conversations Introducing Poetry: chiefly on subjects of natural history for the use of children and young persons* (London: J. Johnson, 1804), 2: 52.
2. *Annual Review, and History of Literature; for 1804* (London: Longman, Hurst, Rees, and Orme, 1805), p. 900.

3. Ibid. The quotation derives from Milton's "Comus," line 222.

4. *Annual Review*, p. 897.

5. Ibid., p. 898.

6. "Cloud," in *The Cyclopædia; or, Universal Dictionary of Arts, Sciences, and Literature*, ed. Abraham Rees et al., vol. 8 (London: Longman, Hurst, Rees, Orme, and Brown, 1802–20).

7. *Athenæum: A Magazine of Literary and Miscellaneous Information* (1807), 1: 4. [Prospectus.]

8. Arthur Aikin, *Journal of a Tour through North Wales and Part of Shropshire, with Observations in Mineralogy, and other branches of Natural History* (London: J. Johnson, 1797).

9. *Athenæum* (1807), 2: 183.

10. *Athenæum* (1809), 5: 539.

11. Luke Howard, *The Climate of London, deduced from Meteorological Observations, made at different places in the Neighbourhood of the Metropolis* (London: W. Phillips, 1818, 1820), 1 (1818): xxxii.

12. Luke Howard, "The Natural History of Clouds," *Journal of Natural Philosophy, Chemistry, and The Arts* 30 (1812): 35–62.

13. *Journal of Natural Philosophy, Chemistry, and The Arts* 30 (1812): 65.

14. *Annals of Philosophy; or, Magazine of Chemistry, Mineralogy, Mechanics, Natural History, Agriculture and the Arts* 1 (1813): 80, 160.

15. *The British Review, and London Critical Journal* 27 (1821): 337–61. See also J. F. Daniell, *Meteorological Essays and Observations* (London: Thomas and George Underwood, 1823), p. 304.

16. Unpublished letter WMS PP/HO/K/A7, Western Manuscripts Collection, Wellcome Institute for the History of Medicine.

17. Unpublished letter WMS PP/HO/K/A1, Western Manuscripts Collection, Wellcome Institute for the History of Medicine.
18. *The Collected Works of Samuel Taylor Coleridge*, ed. H. J. Jackson and J. R. de J. Jackson (London: Routledge, 1971–), 11: 335.
19. *The Poetical Works of William Wordsworth*, ed. Ernest de Selincourt (Oxford: Oxford University Press, 1949), 5: 247.
20. *Gentleman's Magazine* 80 (1810): part 2, p. 631.
21. Ibid., p. 528.
22. *Gentleman's Magazine* 81 (1811): part 2, p. 113.
23. Alan Clark, *Diaries* (London: Weidenfeld and Nicolson, 1993), p. 315.
24. John Bostock, "Remarks upon Meteorology," *Journal of Natural Philosophy, Chemistry, and The Arts* 26 (1810): 9.
25. Ibid., pp. 2–6.
26. Luke Howard, "Observations on Dr. Bostock's Remarks upon Meteorology," *Journal of Natural Philosophy* 26 (1810): 213–4.
27. Ibid., pp. 214–6.
28. John Bostock, "On Meteorological Nomenclature, in answer to Luke Howard, Esq.," *Journal of Natural Philosophy* 26 (1810): 310–11.
29. Thomas Forster, "Specimen of a new Nomenclature for Meteorological Science," *Gentleman's Magazine* 86 (1816): part 1, pp. 131–2.
30. *Gentleman's Magazine* 81 (1811): part 2, p. 113.

31. Thomas Forster, *Researches about Atmospheric Phænomena*, 2d ed. (London: Baldwin, Cradock and Joy, 1815), p. 1.

32. Forster, pp. vi–xiv.

33. *Supplement to the fourth, fifth, and sixth editions of the Encyclopædia Britannica, with Preliminary Dissertations on the History of the Sciences* (Edinburgh: Constable, 1824), 3: 201.

34. Luke Howard, *Climate of London*, 1: xxxii–xxxiii.

35. Thomas Forster, *Researches about Atmospheric Phænomena*, 3d ed. (London: Harding, Mavor, and Lepard, 1823); Forster, *The Perennial Calendar, and Companion to the Almanack; Illustrating the Events of Every Day in the Year, as connected with History, Chronology, Botany, Natural History, Astronomy, Popular Customs, & Antiquities, with Useful Rules of Health, Observations on the Weather; Explanations of the Fasts and Festivals of the Church, and Other Miscellaneous Useful Information* (London: Harding, Mavor, and Lepard, 1824).

36. Forster, *Perennial Calendar*, pp. 93–4.

37. W. H. Smyth, *The Sailor's Word-Book: An Alphabetical Digest of Nautical Terms, including some more especially military and scientific, but useful to seamen; as well as archaisms of early voyagers, etc.* (London: Blackie and Son, 1867).

38. Thomas Milner, *The Gallery of Nature: A Pictorial & Descriptive Tour through Creation* (London: W. S. Orr and Company, 1846), pp. 463–4 (with thanks to Jo Lynch for spotting this one).

39. Henry Stephens, *The Book of the Farm, Detailing the Labours of the Farmer, Farm-Steward, Ploughman, Shepherd, Hedger, Cattle-Man, Field-Worker, and Dairy-Maid* (Edinburgh and London: William Blackwood, 1844), 1: 246.

40. The best account of Forster's and Howard's later membership in the MSL can be found in Nicholas Webb, "Representations of the Seasons in Early-Nineteenth-Century England" (Ph.D. thesis, University of York, 1998).

41. "Sur les modifications des Nuages, et sur les principes de leur production, suspension, et destruction. Extrait d'un Essai lu à la Société Askesienne de Londres en 1803. Par Luke Howard Esqr.," trans. M.-A. Pictet, *Bibliothèque Britannique; ou Recueil Extrait des Ouvrages Anglais périodiques et autres; des Mémoires et Transactions des Sociétés et Académies de la Grande-Bretagne, d'Asie, d'Afrique et d'Amerique*, Sciences et Arts 27 (1804): 185–208.

42. "Ueber die Modificationen der Wolken, von Lucas Howard, Esq. (Ausgez. aus einer zu London im Jahr 1803 gehaltenen Vorles., mit einigen Zusätzen von Pictet)," *Annalen der Physik* 21 (1805): 137–59.

43. "Versuch einer Naturgeschichte und Physik der Wolken, von Lukas Howard, Esq., zu Plaistow bei London," trans. L. W. Gilbert, *Annalen der Physik* 51 (1815): 1–48 and 5–6 (with thanks to James Mackenzie for helping with the German).

44. Cited in *Quarterly Journal of the Royal Meteorological Society* 13 (1887): 163.

45. *American Journal of Science and Arts* 4 (1822): 336.

46. "Cloud," in *Encyclopædia Americana: A Popular Dictionary of Arts, Sciences, Literature, History, Politics and Biography &c.*, ed. Francis Lieber (Philadelphia: Carey, Lea & Carey, 1829–48), 3: 265–6. The second *Encyclopædia Americana*, of 1885, featured a lengthier article, written by Charles Morris of the Philadelphia Academy of Natural Sciences, which discussed the work of Howard, "an able En-

glish meteorologist," in some depth: see "Clouds," in *Encyclopædia Americana: A Supplemental Dictionary of Arts, Sciences, and General Literature* (New York: J. M. Stoddart, 1883–9), 2: 154–6.

47. *American Journal of Science and Arts* 24 (1833): 362.
48. C. S. Rafinesque, *The Good Book, and Amenities of Nature; or Annals of Historical and Natural Sciences* (Philadelphia: Eleutherium of Knowledge, 1840), p. 6.
49. "Notice of the Botanical Writings of the late C. S. Rafinesque," *American Journal of Science and Arts* 40 (1841): 241.
50. Elias Loomis, "Meteorological Observations made at Hudson, Ohio," *American Journal of Science and Arts* 41 (1841): 310–31; chart, p. 325.
51. Reproduced in James Roger Fleming, *Meteorology in America, 1800–1870* (Baltimore and London: Johns Hopkins University Press, 1990), pp. 84–5.

9. Fame

1. Luke Howard, *My Ledger; or, A compromise with prudence. Written in 1808* (London: Taylor and Francis, 1856), pp. 12–13.
2. John Claridge, *The Shepherd of Banbury's Rules To judge of the Changes of the Weather, Grounded on Forty Years Experience; To which is added, A rational Account of the Causes of such Alterations, the Nature of Wind, Rain, Snow, &c. on the Principles of the* Newtonian *Philosophy* (London: W. Bickerton, 1744), p. ii. The book went through twelve editions.

3. Ibid.

4. Luke Howard, *On the Modification of Clouds, &c.* (London: J. Taylor, 1804), p. 4.

5. *Gentleman's Magazine*, 18 (1748), p. 255.

6. Ibid.

7. James Clerk Maxwell, "Molecular Evolution," in *Poems of Science*, ed. John Heath-Stubbs and Phillips Salman (Harmondsworth: Penguin, 1984), p. 231.

8. *Life of William Allen, with Selections from his Correspondence* (London: Charles Gilpin, 1847), 1: 68. Like Howard, Allen has had a street named after him in London.

9. Ibid., 1: 57–61.

10. Gough's little Spaniel, Music, had sat by the body of her master for three lonely months until discovered by a passing herdsman. See H. D. Rawnsley, "The Story of Gough and His Dog," in *Past and Present in the English Lakes* (Glasgow: MacLehose and Sons, 1916), pp. 153–208.

11. *Life of William Allen*, 1: 87.

12. Luke Howard, "Journey in Westmorland 1807," London Metropolitan Archives, Acc. 1017/1397.

13. James Plumptre, *The Lakers: A Comic Opera, in Three Acts* (London: W. Clarke, 1798), pp. 6, 40.

14. Richard Holmes, *Coleridge: Early Visions* (London: Hodder and Stoughton, 1989), p. 328.

15. Cited in Malcolm Andrews, *The Search for the Picturesque: Landscape Aesthetics and Tourism in Britain, 1760–1800* (Aldershot: Scolar Press, 1989), p. 233.

16. Holmes, p. 278.

17. Mary Shelley, *Frankenstein* (Harmondsworth: Penguin Classics, 1984), p. 244.
18. See Marjorie Hope Nicolson, *Newton Demands the Muse: Newton's Opticks and the Eighteenth Century Poets* (Princeton, N.J.: Princeton University Press, 1946), pp. 1–2.
19. *The Diary of Benjamin Robert Haydon*, ed. Willard Bissell Pope (Cambridge, Mass.: Harvard University Press, 1960) 2: 173–76.
20. Timothy Hilton, *Keats and his world* (London: Thames & Hudson, 1971), p. 62.
21. John Keats, *Complete Poems,* ed. Jack Stillinger (Cambridge, Mass.: Harvard University Press, Belknap Press, 1978), p. 357.
22. James Thomson, *A Poem Sacred to the Memory of Sir Isaac Newton* (London: J. Millan, 1727), p. 10.
23. Cited in Christopher Lawrence, "The power and the glory: Humphry Davy and Romanticism," in *Romanticism and the Sciences,* ed. Andrew Cunningham and Nicholas Jardine (Cambridge: Cambridge University Press, 1990), p. 221.
24. Euan Nisbet, "In Retrospect," *Nature* 388 (10 July 1997): 137.

10. The Beaufort Scale

1. Robert FitzRoy, *The Weather Book: A Manual of Practical Meteorology* (London: Longman, 1863), p. 31.
2. William Scoresby, Jr., *An Account of the Arctic Regions, with a History and Description of the Northern Whale-Fishery* (Edinburgh: Constable, 1820), p. 250.
3. William Scoresby, Jr., *Meteorological Observations on a Greenland Voyage, in the Ship Resolution in the Year 1810* (1810).

4. Scoresby, *Account of the Arctic Regions*, pp. 419–20.
5. Fergus Fleming, *Barrow's Boys* (London: Granta Books, 1998), p. 31.
6. Tom and Cordelia Stamp, *William Scoresby: Arctic Scientist* (Whitby: Caedmon, 1976), p. 68.
7. Scoresby, *Account of the Arctic Regions*, p. 396.
8. William Falconer, "Wind," in *An Universal Dictionary of the Marine: or, a copious explanation of the Technical Terms and Phrases employed in the Construction, Equipment, Furniture, Machinery, Movements, and Military Operations of a Ship* (London: T. Cadell, 1769).
9. Scoresby, *Account of the Arctic Regions*, p. 396.
10. Daniel Defoe, *The Storm: or, a Collection of the most remarkable Casualties and Disasters which happen'd in the Late Dreadful Tempest, both by Sea and Land* (London: G. Sawbridge, 1704), pp. 21–2. Also cited in L. G. Garbett, "Admiral Sir Francis Beaufort and the Beaufort Scales of Wind and Weather," *Quarterly Journal of the Royal Meteorological Society* 52 (1926): 164.
11. Falconer, "Wind."
12. Alfred Friendly, *Beaufort of the Admiralty: The Life of Sir Francis Beaufort 1774–1857* (London: Hutchinson, 1977), p. 129.
13. Ibid., p. 142.
14. Ibid., p. 144.
15. Ibid., p. 145.
16. H. T. Fry, "The Emergence of the Beaufort Scale," *Mariner's Mirror* 53 (1967): 311–13 and 54 (1968): 412.
17. Fry, "Emergence of the Beaufort Scale," p. 311.

18. William Burney, "Breeze," in *A New Universal Dictionary of the Marine, etc.* (London: T. Cadell and W. Davies, 1815), p. 57.
19. Friendly, p. 144.
20. *The Autobiography of Charles Darwin 1809–1882*, ed. Nora Barlow (London: Collins, 1958), p. 73.
21. Friendly, p. 146.
22. "The Log-Board," *Nautical Magazine* 1 (1832): 537–8. Also in Friendly, p. 146.
23. Ibid., p. 538.
24. Blair Kinsman, "Historical Notes on the Original Beaufort Scale," *Marine Observer* 39 (1969): 124.
25. Bram Stoker, *Dracula* (Oxford: World's Classics, 1996), p. 75 (with thanks to Markman Ellis for pointing it out).
26. Lyall Watson, *Heaven's Breath: A Natural History of the Wind* (London: Hodder and Stoughton, 1984), p. 213.

II. Goethe and Constable

1. From J. W. von Goethe, "Atmosphere," cited in Kurt Badt, *John Constable's Clouds* (London: Routledge and Kegan Paul, 1950), p. 13.
2. Reproduced in Elizabeth Fox Howard, "Goethe and Luke Howard, F.R.S.," *Friends Quarterly Examiner* 66 (1932): 224–5.

3. A. W. Slater, "Luke Howard, F.R.S. (1772–1864) and his Relations with Goethe," *Notes and Records of the Royal Society* 27 (1972): 122.
4. J. W. von Goethe, *Italian Journey 1786–1788*, trans. W. H. Auden and Elizabeth Mayer (Harmondsworth: Penguin, 1970), p. 23.
5. Ibid., p. 27.
6. Ibid.
7. Ibid., p. 32.
8. Ibid., pp. 31–2.
9. Cited in Badt, *Constable's Clouds*, p. 18.
10. This contemporary translation (perhaps by his son Robert) is taken from a notebook, copied out in Howard's hand, now in the London Metropolitan Archives: Acc. 1017/1517.
11. J. W. von Goethe, "Wolkengestalt nach Howard," *Zur Naturwissenschaft überhaupt* (Stuttgart and Tübingen, 1817–23), 1: 97–125.
12. Goethe, "Wolkengestalt nach Howard," pp. 124–5; and in D.F.S. Scott, *Some English Correspondents of Goethe* (London: Methuen, 1949), pp. 51–4.
13. Scott, *Some English Correspondents*, pp. 48–54; Gold's *London Magazine and Theatrical Inquisitor* 4 (1821). The dual-language version, complete with commentary, was subsequently republished in the German periodical where the original four cloud poems had appeared in 1817: Goethe, pp. 322–3.
14. Scott, *Some English Correspondents*, pp. 53–4; Goethe, pp. 326–9.

15. Percy Bysshe Shelley, "The Cloud." First published in *Prometheus Unbound: A Lyrical Drama in four acts, with other poems* (London: C. and J. Ollier, 1820), pp. 196–200.

16. For a fuller interpretation, see Desmond King-Hele, *Shelley: His Thought and Work,* 2d ed. (London: Macmillan, 1970), pp. 219–27; F. H. Ludlam, "The Meteorology of the Ode to the West Wind," *Weather* 27 (1972): 503–14; and J. E. Thornes, "Luke Howard's Influence on Art and Literature in the Early Nineteenth Century," *Weather* 39 (1984): 254.

17. *Mégha Dúta; or Cloud Messenger; a poem, in the Sanscrit Language,* trans. Horace Hayman Wilson (Calcutta: Hindoostanee Press, 1813), p. 32 (with thanks to Nigel Leask for the reference).

18. Percy Bysshe Shelley, *The Cloud-Die Wolke,* translated into German verse by "P. H." (privately printed, c. 1830).

19. T. J. Reed, *Goethe* (Oxford: Oxford University Press, 1984), p. 44.

20. Cited in Timothy Wilcox, "Keeping Time: Clouds and Chronometry in Constable's Major Landscapes," in *Constable's Clouds: Paintings and Cloud Studies by John Constable,* ed. Edward Morris (Edinburgh and Liverpool: National Galleries of Scotland and National Museums and Galleries on Merseyside, 2000), p. 163.

21. From the notebook in the LMA: Acc. 1017/1517.

22. "Memoir of Luke Howard," in *Luke Howard (1772–1864) His correspondence with Goethe and his Continental Journey of 1816,* ed. D.F.S. Scott (York: William Sessions, 1976), p. 1.

23. Ibid., p. 4.

24. Ibid., p. 8.

25. Notebook, LMA: Acc. 1017/1517.

26. *Goethe on Science: A Selection of Goethe's Writings,* ed. Jeremy Naydler (Edinburgh: Floris Books, 1996), p. 128.

27. Cited in Badt, *Constable's Clouds*, p. 15, and John E. Thornes, *John Constable's Skies: A Fusion of Art and Science* (University of Birmingham Press, 1999), p. 190.

28. Badt, p. 20.

29. Marie Lødrup Bang, *Johan Christian Dahl 1788–1857: Life and Works* (Oslo: Norwegian University Press, 1987), 1: 199.

30. Joseph Leo Koerner, *Caspar David Friedrich and the subject of landscape* (London: Reaktion Books, 1990), p. 193.

31. Bang, p. 79.

32. *Letters from Goethe*, trans. M. von Herzfeld and C. Melvil Sym (Edinburgh University Press, 1957), p. 520.

33. Cited in Thornes, *Constable's Skies*, p. 21.

34. The highlights are Badt, *Constable's Clouds*; Hubert Damisch, *Théorie du Nuage: pour une histoire de la peinture* (Paris: Editions du Seuil, 1972); Thornes, *Constable's Skies*; and *Constable's Clouds: Paintings and Cloud Studies by John Constable,* ed. Edward Morris (Edinburgh and Liverpool: National Galleries of Scotland and National Museums and Galleries on Merseyside, 2000).

35. Cited in Wilcox, "Keeping Time," p. 169.

36. Cited in E. H. Gombrich, *Art and Illusion: A study in the psychology of pictorial representation* (Oxford: Phaidon, 1960), p. 150.

37. Charles Taylor, *The Landscape Magazine: Containing Preceptive Principles of Landscape* (London: C. Taylor, 1792), p. 105.

38. Alexander Cozens, *A New Method of Assisting the Invention in Drawing Original Compositions of Landscape* (London: for the author, 1785). See also A. P. Oppé, *Alexander & John Robert Cozens* (London: A. and C. Black, 1952), pp. 48–51, and Kim Sloan, *Alexander and John Robert Cozens: The Poetry of Landscape* (New Haven and London: Yale University Press, 1986), pp. 85–6.

39. Thornes, *Constable's Skies*, p. 177.

40. Ibid., p. 250.

41. Ibid., p. 78.

42. Ibid., p. 73.

43. Ibid., p. 57.

44. Ibid., p. 51.

45. C. R. Leslie, *The Life of John Constable, composed chiefly of his letters* (London: Phaidon, 1951), p. 101.

12. The International Year of Clouds

1. Lord Byron, "Don Juan," in *The Complete Poetical Works*, ed. Jerome J. McGann (Oxford: Oxford University Press, 1980–93), 5: 79.

2. *Philosophical Magazine and Journal* 62 (1823): 229.

3. Ibid., p. 305.

4. George J. Symons, "The History of English Meteorological Societies, 1823 to 1880," *Quarterly Journal of the Royal Meteorological Society* 7 (1881): 65–98; Peter R. Cockrell, "The Meteorological Society of London 1823–1873," *Weather* 23 (1968): 357–61; J. M. Walker, "The Meteorological Societies of London," *Weather* 48 (1993): 364–72.
5. Nicholas Webb, "Representations of the Seasons in Early-Nineteenth-Century England," (Ph.D. thesis, University of York, 1998), p. 127; Walker, "Meteorological Societies of London," p. 366.
6. See George J. Symons, "History of English Meteorological Societies," p. 70; Webb, pp. 128–31.
7. John Galt, *Bogle Corbet; or, The Emigrants* (London: Colburn & Bentley, 1831), 1: 264.
8. *A Catalogue of Cabinet and other Esteemed Paintings, Prints, Coins, Medals, Philosophical and Mathematical Instruments, Stained Glass, Curious Chinese Furniture, large Ornamental Jars &c &c. of Alexander Tilloch* (London: Robert Saunders, 1825).
9. Ralph Abercromby, "On the Identity of Cloud Forms all over the World," *Quarterly Journal of the Royal Meteorological Society* 13 (1887): 141.
10. See J. A. Kington, "A Century of Cloud Classification," *Weather* 24 (1969): 84–9.
11. *The Friend* (1864), p. 100.
12. *Short Memorials of the late Luke and Mariabella Howard, of Ackworth Villa, Yorkshire,* by an aged relative (Tottenham: privately printed, 1864), p. 13.
13. *The Friend* (1864), p. 100.

14. Cited in Michael Wolfers, "A Head in the Clouds," *Illustrated London News* (February 1973), p. 53.

15. H. Hildebrand Hildebrandsson, *Sur la Classification des Nuages employée à l'Observatoire Météorologique d' Upsala*, 2d ed. (Uppsala: C. J. Lundström, 1880), p. 1.

16. Ralph Abercromby, "Suggestions for an International Nomenclature of Clouds," *Quarterly Journal of the Royal Meteorological Society* 13 (1887): 155.

17. Ibid., pp. 154–66.

18. H. H. Hildebrandsson, W. Köppen, and G. Neumayer, *Wolken-Atlas.—Atlas des Nuages.—Cloud Atlas.—Moln-Atlas*. (Hamburg: Gustav W. Seitz, 1890).

19. International Meteorological Committee, *Atlas International des Nuages/International Cloud Atlas/Internationaler Wolken-Atlas* (Paris: Gauthier-Villars et Fils, 1896), p. 11.

20. *Quarterly Journal of the Royal Meteorological Society* 13 (1887): 154.

21. Ibid., p. 146.

22. *American Meteorological Journal: A Monthly Review of Meteorology, Medical Climatology, and Geography* 8 (1891–92): 526.

23. IMC, *Cloud Atlas*, p. 13.

24. *Symons's Monthly Meteorological Magazine* 31 (1896): 82.

Epilogue: Afterlife

1. *The Friend: A Religious, Literary and Miscellaneous Journal,* new series 4 (1864): 100.
2. Bernard Howard, "A Luke Howard Miscellany: Compiled by his Great Grandson" (London: unpublished typescript, 1959), pp. 138–9.
3. Nicholas Webb, "Representations of the Seasons in Early-Nineteenth-Century England" (Ph.D. thesis, University of York, 1998), p. 146.
4. *Quarterly Journal of the Royal Meteorological Society* 23 (1897): 62.
5. *Quarterly Journal of the Royal Meteorological Society* 13 (1887): 163.

Acknowledgments

For their wise counsel throughout the various stages of this book's completion, I would like to thank Joanna Lynch, Markman Ellis, Gavin Jones, Alexa de Ferranti, Piers Russell-Cobb, and Peter Straus. Among the family, friends, and colleagues who have shared their kindness and conversation, I would also like to thank Peter de Bolla, Chlöe Chard, Susan Coleridge, Justin Croft, Elizabeth Eger, Angela Foster, Dan Franklin, Charlotte Grant, Jennifer Greitschus, Michael Griffiths, David Hamblyn, Dorothy Hamblyn, Judith Hawley, Claudia Jessop, Nigel Leask, Anthony, Paula, and Emma Lynch, James Mackenzie, Steve McLeish, Sean O'Connor, Felix Pryor, Simon Schaffer, Becky Senior, John Shaw, Nicholas Webb, and Peter Wilson.

It is also a pleasure to thank the staff of the libraries, institutions, and archives whose efforts so materially assisted the research and writing of this book: the British Library; Cambridge University Library; Friends' House Library, Euston Road, London; Glaxo Wellcome plc, Greenford, U.K.; London Metropolitan Archives; the Meteorological Office; Newham Local History Library, Stratford; the Royal Meteorological Society; the Science Museum; the Tate Gallery, Millbank; University of London Library; and the Western Manuscripts Department of the Wellcome Institute for the History of Medicine. I would like to extend special thanks to Jude White, Warden of the Winchmore Hill Friends' Meeting House, who kindly showed me around the gardens where Luke Howard's body has lain, unmarked, for the last one hundred and forty years.

Index